MANUFACTURING SCIENCE

AF656452

By :

Vikas Upadhyay

B.E. (Hons.), M.Tech.

Vikas Agarwal

B.Tech., M.Tech.

Department of Mechanical Engineering
Faculty of Engineering and Technology
R.B.S. College, Agra

S.K. KATARIA & SONS®

Publisher of Engineering Books

4885/109, Prakash Mahal, Dr. Subhash Bhargav Lane,
Opposite Delhi Medical Association, Daryaganj, New Delhi–110002
Phone: +91-11-43551243; Mobile: +91-9871775858, 7678425209
e-mail: katariabook@gmail.com, katariabooks@yahoo.com
Website: www.skkatariaandsons.com

(e-books, subscriptions and bulk buying also available on our website)

Published by:

S.K. Kataria & Sons®

4885/109, Prakash Mahal, Dr. Subhash Bhargav Lane,
Opposite Delhi Medical Association, Daryaganj, New Delhi-110002
Phone: +91-11-43551243; Mobile: +91-9871775858, 7678425209
e-mail: katariabook@gmail.com, katariabooks@yahoo.com

First Edition: August, 2012; Reprint: 2026

***Price :* ₹ 295/-**

Laser Type Setting by:
SARAASSIGNMENTS, Delhi

Printed at :
A.S. OFFSET PRESS, Delhi

PREFACE

The book 'Manufacturing Science' is basically a revised and extended version of earlier book entitled 'Manufacturing Processes'. Some topics are added while some are removed from the previous book to make it strictly as per university syllabus.

We sincerely believe that the book 'Manufacturing Science' will remain useful and suitable to our readers as the earlier book. We welcome constructive comments, suggestions and criticisms from our students as well as faculty members to improve upon the book contents in future.

Vikas Upadhyay
Vikas Agarwal

Dedicated to our Parents for their Love and Untiring Support

CONTENTS

UNIT-I

Properties: Inspection and Testing of Material

1. ENGINEERING MATERIALS

SYLLABUS

Properties, Inspection and Testing of Materials: Introduction to Stress and Strain

Mechanical Properties : Strength, Elasticity, Stiffness, Malleability, Ductility, Brittleness, Toughness and Hardness. Elementary Ideas of Fracture, Fatigue and Creep.

Testing of Metals : Destructive Testing, tensile testing, Compression test, Hardness tests, Impact test

CHAPTER 1

Engineering Materials

INSIDE THIS CHAPTER

1.1. INTRODUCTION

Materials are one of the main component of manufacturing system and are selected on the basis of desired properties and characteristics to perform the specific task with minimum cost. So, the properties and application of materials are of great importance for engineers to select a proper material for practical use under given conditions.

1.2. CLASSIFICATION

Materials comprise a wide range of metals and non-metals, and can be broadly classified on the basis of origin and composition as :

1. Metals and Alloys
2. Non-metals.

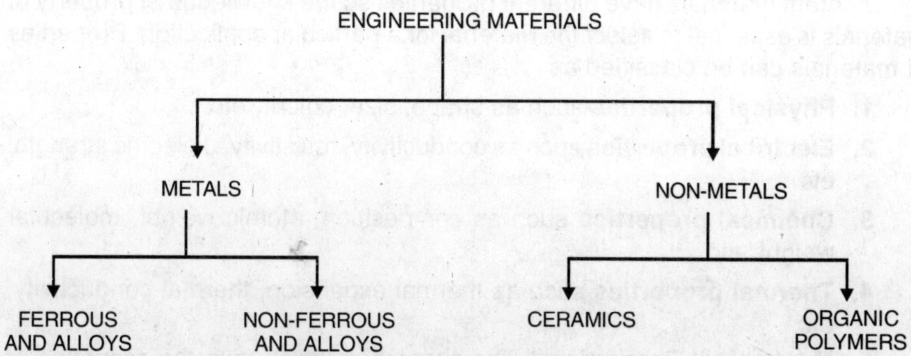

Fig. 1.1. Classification of Engineering Materials.

1.3. METALS AND ALLOYS

Metals are generally solids and have good heat and electrical conductivity, good malleability and ductility, high melting point and high strength. These can be further classified as :

1. Ferrous metal and alloys
2. Non ferrous metal and alloys

1.3.1. Ferrous Metal and Alloys

Ferrous materials contain iron as their main constituent and are most widely used because of their low cost and varied mechanical properties. Examples are cast iron, pig iron, wrought iron, steel, etc.

1.3.2. Non Ferrous Metal and Alloys

Non ferrous materials are those which contain a metal other than iron as their main constituent. Examples are aluminium, copper, zinc, etc.

1.4. NON-METALS

Non metals are light in weight, possess poor heat and electrical conductivity, poor ductility and low strength. These can be further classified as :

1. Ceramics
2. Organic Polymers

1.4.1. Ceramics

Ceramics are basic oxides and are compounds of metallic and non-metallic elements. Examples are concrete, ferrites, refractories, silicon carbides, etc.

1.4.2. Organic Polymers

These materials are carbon compounds which have high molecular weight and complex three dimensional structures due to long molecular chains. Examples are PVC, polyethylene, rubber, cotton, etc.

1.5. PROPERTIES OF MATERIAL

Different materials have different properties so the knowledge of property of materials is essential to select the material for a particular application. Properties of materials can be classified as

1. **Physical properties** such as shape, size, colour, etc.
2. **Electrical properties** such as conductivity, resistivity, dielectric strength, etc.
3. **Chemical properties** such as composition, atomic weight, molecular weight, etc.
4. **Thermal properties** such as thermal expansion, thermal conductivity, etc.
5. **Mechanical Properties :** The properties which give the response of the material under externally applied forces are known as mechanical properties. Material subjected to forces may have elastic deformation,

plastic deformation or fracture depending upon the magnitude of load. The knowledge of mechanical properties is very crucial for a design engineer to design the machines and structures. These properties determine the usefulness of the material under given conditions to perform its intended operation without failure.

Important mechanical properties of engineering materials are

(i) Strength
(ii) Elasticity
(iii) Plasticity
(iv) Stiffness
(v) Ductility
(vi) Malleability
(vii) Brittleness
(viii) Toughness
(ix) Resilience
(x) Hardness

1.5.1. Strength [UPTU May 2009]

The resistance offered by a material when subjected to external load is known as strength. Higher the strength the higher is the capacity of the material to withstand load without rupture or failure. Depending on the type of load the strength can be tensile, compressive, shear, bending or torsional.

Resistance offered by the material is stress, so the strength can be measured in term of stress. The stress required to cause permanent deformation is called yield strength and the maximum stress before fracture is called ultimate strength.

$$\text{Stress (tensile, shear, compressive, bending or torsional)} = \frac{\text{Load (tensile, shear, compressive, bending or torsional)}}{\text{Initial cross-sectional area}}$$

1.5.2. Elasticity

Elasticity is defined as the property of the material due to which it is able to regain its original shape and size when the external load is removed. Actually no material is perfectly elastic over entire range i.e. they are elastic up to a certain value of stress known as elastic limit. The part of the elastic limit within which stress is directly proportional to strain is known as proportional limit.

1.5.3. Plasticity

Plasticity is defined as the property of the material which enables permanent deformation of material without failure or rupture. Processes such as forming, rolling, extrusion, etc. require plasticity in materials.

1.5.4. Stiffness [UPTU Dec. 2008]

Stiffness is defined as the property of the material due to which it resists deflection or deformation. It is also known as rigidity. The degree of stiffness of the material is represented by Modulus of Elasticity. Small value of modulus of Elasticity represents flexible material and large value represents stiff and rigid material.

1.5.5. Ductility [UPTU Dec. 2008]

Ductility is a property of material by virtue of which it can be drawn into wires without rupture and without loosing much strength. It is measured by the percentage elongation or percentage reduction in the cross sectional area before rupture.

$$\text{Percentage elongation} = \frac{\text{Increase in length}}{\text{Initial length}} \times 100$$

$$\text{Percentage reduction in area} = \frac{\text{Initial area} - \text{Final area}}{\text{Initial area}} \times 100$$

Metals having percentage elongation more than 15% are ductile. Some examples of ductile material are mild steel, copper, aluminium, etc.

1.5.6. Malleability [UPTU Dec. 2008, May 2009]

Malleability is defined as the property of material due to which it can be converted into thin flat sheet without cracking. Gold, silver, tin and lead are examples of metals exhibiting high malleability. In malleability plastic deformation refers particularly to compressive loading whereas deformation in ductility refers to tensile loading. Lead has high malleability but low ductility.

1.5.7. Brittleness [UPTU Dec. 2008]

Brittleness is the property of a material by virtue of which it fails or fractures without any appreciable deformation. Brittleness shows the lack of plasticity and should not be considered as lack of strength. It is opposite to ductility. Cast iron, glass, concrete, etc. shows brittle fracture i.e. a very small amount of plastic deformation before fracture.

1.5.8. Toughness [UPTU Dec. 2008, May 2009]

Toughness is defined as the property of material which refers to the strain energy absorbing capacity of the material before its failure. It is equal to work per unit volume needed to fracture the material. Toughness decreases with the rise in temperature. This property plays a vital role in the design of components subjected to shock or impact loading e.g. power press punch and pneumatic hammer.

1.5.9. Resilience

Resilience is the ability of material to absorb energy within elastic limit. This energy is released when external force is removed. Resilience is equal to the amount of energy absorbed per unit volume within elastic limit and indicates the ability of material to withstand shock and vibrations.

1.5.10. Hardness

Hardness is defined as the ability of a material to resist abrasion, indentation, wear and scratch. Increase in hardness reduces the machinability of the material due to decrease in toughness.

Most commonly used test to determine the hardness of materials are Brinell, Rockwell and Vickers hardness test. In these tests, indenters are used to produce indentation on the surface and the relative hardness is given in numbers.

1.6. STRESS AND STRAIN

When some external force is applied on the body, it undergoes deformation. The molecule of body sets up some resistance to this deformation. This resistance of material per unit area is called stress. The term stress is also called as mean intensity of stress. Mathematically, it is given by the following relation

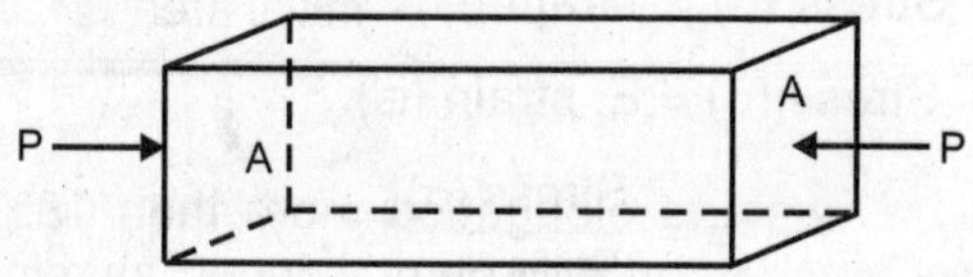

Fig. 1.2

$$\sigma = \frac{P}{A}$$

where σ is stress, N/m^2

P is load or force acting on the body, N

A is cross-section area of body, m^2

Nominal Stress : The ratio of load at fracture to the original cross-sectional area is called nominal stress i.e.,

$$\text{Nominal stress} = \frac{\text{Load at fracture}}{\text{Initial cross-sectional area}}$$

True Stress : The ratio of load at fracture to final cross-sectional area is called true stress i.e.,

$$\text{True stress} = \frac{\text{Load at fracture}}{\text{Final cross-sectional area}}.$$

When external force is applied on a body, it undergoes deformation. This deformation per unit length is termed as strain. Mathematically it is given by

$$\epsilon = \frac{\delta l}{l}$$

where ϵ is strain

δl is change in length of body

l is original length of the body

Strain is a dimensionless quantity.

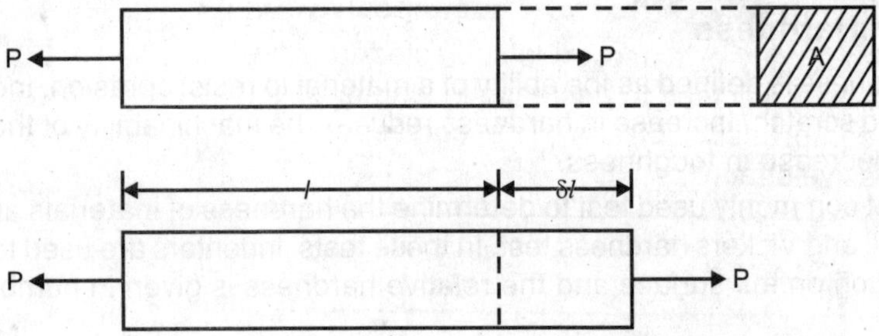

Fig. 1.3

Relation Between Stress and Strain (Hooke's Law)

Hooke's law states that "within elastic limit, stress is directly proportional to strain." Mathematically, the relation is given by

$$\text{Stress } (\sigma) \propto \text{strain } (\in)$$

or
$$\text{Stress } (\sigma) = \text{E. strain } (\in)$$

or
$$E = \frac{\text{Stress } (\sigma)}{\text{Strain } (\in)}$$

Where, E is a constant of proportionality known as Young's modulus or modulus of elasticity of the material. The modulus of elasticity is the slope of the stress-strain diagram in the linear elastic region and its value depends upon the particular material being used. The units of E is same as the units of stress, because strain is a dimensionless quantity.

Types of Stress and Strain :

1. Tensile stress and tensile strain

When a body or a section is subjected to two equal and opposite pulls, then the body will tend to increase in length and the stress thus induced is called tensile stress. The corresponding strain induced in the body is called tensile strain.

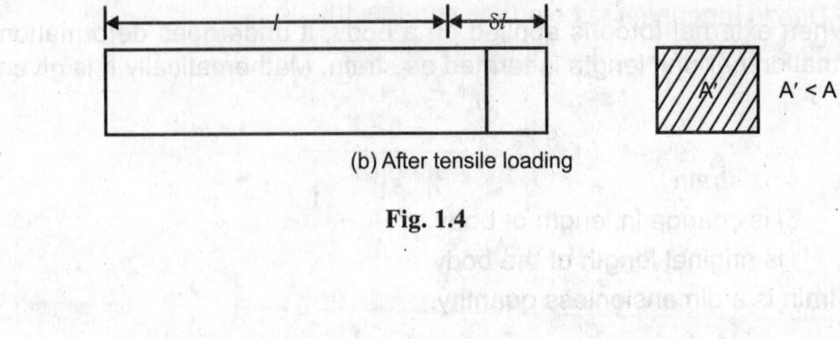

Fig. 1.4

2. Compressive stress and compressive strain

When a body or a section is subjected to two equal and opposite pushes, then the body will tend to decrease in length, as a result, the stress induced is called compressive stress. The corresponding strain induced is called compressive strain.

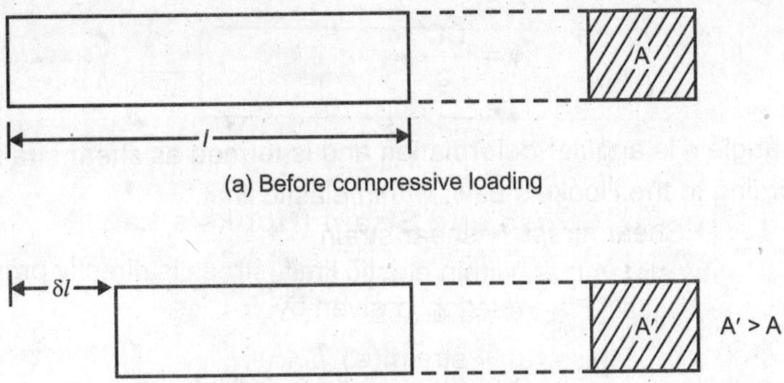

(a) Before compressive loading

(b) After compressive loading

Fig. 1.5

3. Shear stress (τ) and Shear strain (ϕ)

When a force is applied tangentially on the surface of a body (as shown in Fig. 1.6), the stress induced in the body is called shear stress and the corresponding strain is called shear strain. Shear stress is denoted by τ and shear strain is denoted by ϕ.

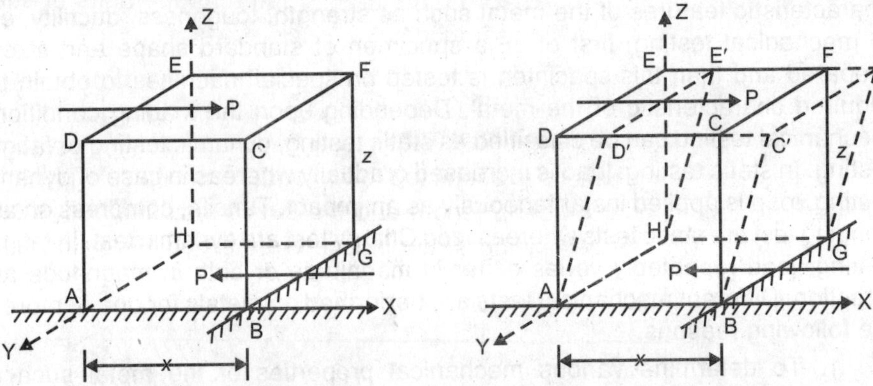

(i) Before shearing (ii) After shearing

Fig. 1.6

Let us consider a cuboid ABCDEFGH (as shown in Fig. 1.6) having ABCD plane fixed at bottom.

Due to tangential force on the surface, the body gets distorted and take new shape as ABC′D′E′F′GH due to the deformation, then from Fig. 1.6.

$$\text{Shear stress, } \tau = \frac{P}{\text{Area of DCFE}}$$

or

$$\tau = \frac{P}{xy}$$

From the $\Delta BCC'$,

$$\tan \phi = \frac{CC'}{BC}$$

ϕ is very small, then

$$\phi = \frac{CC'}{z}$$

This angle ϕ is angular deformation and is termed as shear strain.

According to the Hooke's Law, within elastic limit,

Shear stress $\propto$ shear strain

$$\therefore \qquad \tau = C.\phi$$

$$\text{or} \qquad C = \frac{\tau}{\phi}$$

Where, C is constant of proportionality, called modulus of rigidity of the material.

1.7. TESTING OF METALS

Testing of metals is essential to obtain the knowledge regarding the various characteristic features of the metal such as strength, toughness, ductility, etc. In mechanical testing, first of all a specimen of standard shape and size is prepared and then this specimen is tested on special machines to obtain the required characteristic of the metal. Depending upon the loading conditions, mechanical testing can be classified as static testing, dynamic testing or fatigue testing. In static testing, load is increased gradually whereas in case of dynamic testing, load is applied instantaneously as an impact. Tensile, compression and bending test are static tests whereas Izod Charpy test are dynamic test. In fatigue testing, load repeatedly varies either in magnitude or both in magnitude and direction. Different mechanical tests are performed on metals for one or more of the following reasons.

1. To determine various mechanical properties of the metal such as strength, toughness, ductility, malleability, etc.
2. To provide specific data to the design engineer so that he can take decision that whether the particular material can be used for required application or not. Moreover, it also helps the designer in various stages of designing a component.
3. To check the suitability of the metal under various operating conditions.

Depending upon the condition of the specimen after the test, testing methods can be divided into destructive test and non destructive test. In destructive testing, the test specimen is completely damaged and therefore new specimen is required for each test. Tensile test, compression test, impact test, fatigue test, creep test are destructive test methods. In non destructive testing, testing is performed in such a way that it does not cause any damage to the specimen and hence maintains the usability of the specimen for further tests/applications. Visual examination radiography test, magnetic test, ultrasonic test are non destructive tests.

1.7.1 Destructive Tests

(i) **Tensile test :** Tensile test is performed to find out different properties such as ultimate tensile strength, modulus of elasticity, yield strength, elastic limit, proportional limit, breaking strength etc.

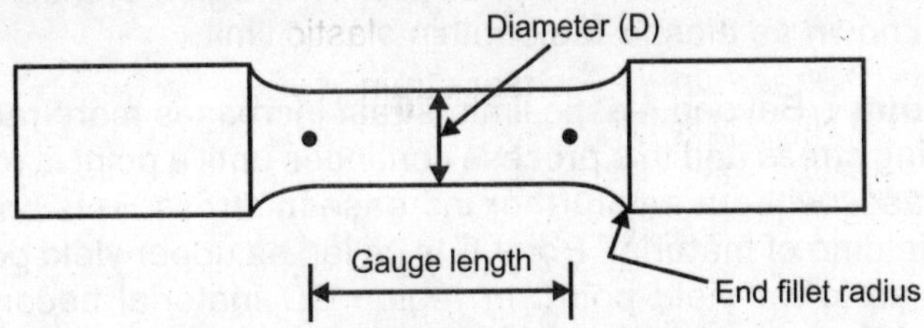

Fig. 1.7 Tensile testing specimen.

The tensile strength of material is determined by gripping the specimen between the jaws of universal testing machine and simultaneously measuring the load and elongation in a specified length, called as gauge length. Results thus obtained depend upon the size of specimen and can not be applied to member of any size. So to apply the results to members of any size the axial load (P) is divided by cross section area (A) to obtain load per unit area known as stress (σ) and elongation (δl) is divided by length (l) to obtain unit elongation known as strain ($\in$). These data are then plotted on a graph with strain as abscissa and stress as ordinate to form stress strain curve.

Stress strain diagram for a ductile material say mild steel is shown in Fig. 1.8. Salient features of the diagram are discussed below.

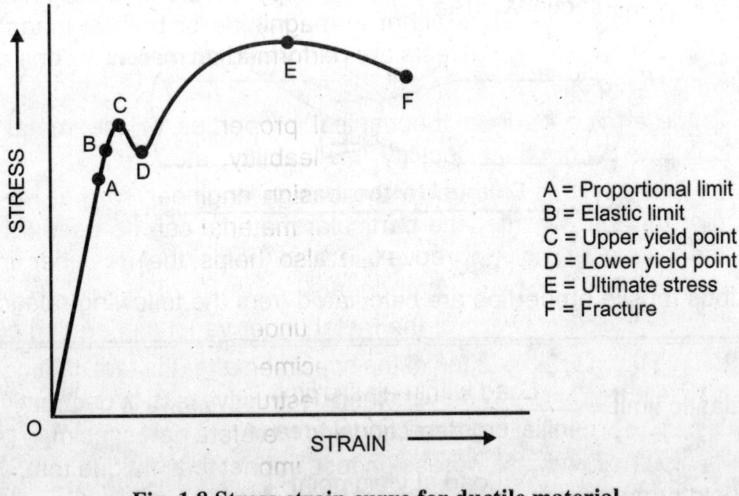

Fig. 1.8 Stress-strain curve for ductile material

Proportional limit : In the beginning of the test when load is applied on specimen, strain increases in direct proportion to stress and obeys Hooke's law. The diagram begins with straight line from origin O to point A representing linear and proportional relationship between stress and strain. Beyond point A stress is not proportional to strain; hence the stress at A is called as proportional limit.

Elastic Limit : Beyond the proportional limit, stress is not proportional to strain and depart from straight line relationship. However, the material remains elastic upto point B and returns to its original from when the load is removed. Stress at B is called as elastic limit stress. The region of the curve between O and B is known as elastic region.

Yield Point : Beyond elastic limit, strain increases more rapidly than the corresponding stress and this process continues until a point is reached where strain increases without any further increase in stress. This phenomenon is known as yielding of material. Point C is called as upper yield point and point D is called as lower yield point. In region CD material becomes perfectly plastic and deforms without an increase in the applied load.

Ultimate Strength : After yielding, the material begins to strain harden and an increase in load is required for further elongation. With further increase in load, stress again starts increasing till the stress reaches the maximum value at point E. Stress corresponding to point E is known as ultimate stress of material and corresponds to ultimate strength of material.

Breaking Strength : Beyond point E, neck formation takes place and the apparent stress decreases until fracture takes place at F. Point F is called as fracture or breaking point and the corresponding stress is called as breaking stress.

After fracture the two separated portions of the test piece are placed together as shown in Fig. 1.9 and distance between the gauge marks and diameter at place of fracture are measured to determine the increase in gauge length and decrease in cross-sectional area.

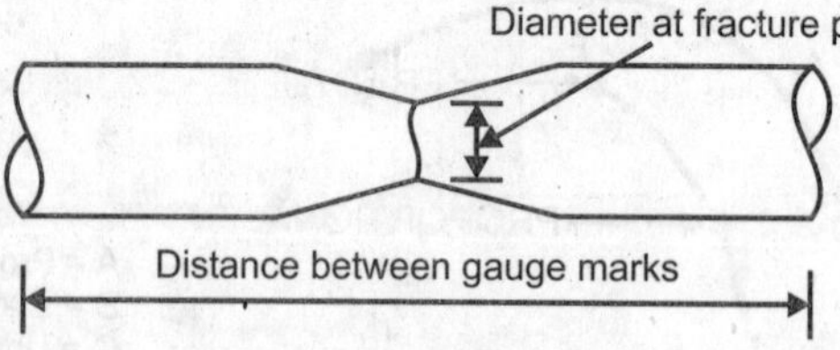

Fig. 1.9

Various tensile properties are calculated from the following relations.

1. Elastic limit $= \dfrac{\text{Load within elastic limit}}{\text{Initial cross-sectional area (A)}}$ where $A = \dfrac{\pi D^2}{4}$

2. Yield strength $= \dfrac{\text{Load at yield point}}{\text{Initial cross-sectional area}}$

3. Ultimate tensile strength = $\frac{\text{Ultimate load}}{\text{Initial cross-sectional area}}$

4. Percentage elongation = $\frac{\text{Increase in gauge length}}{\text{Initial gauge length}} \times 100$

5. Percentage reduction in area = $\frac{\text{Reduction in area}}{\text{Initial cross-sectional area}} \times 100$

6. Young's modulus of elasticity (E) = $\frac{\text{Stress}}{\text{Strain}}$

7. Breaking strength = $\frac{\text{Breaking load}}{\text{Initial cross-sectional area}}$

(ii) **Compression Test :** Compression test are performed for brittle materials like cast iron, ceramics, bricks, concrete etc as they can not be subjected to tensile test. Compression test are also conducted on universal testing machine and are just reverse of tensile test in respect of direction of application of axial load. In compression test the specimen tends to shorten in length on application of load while in tensile test it elongates. In compression testing a circular section is usually preferred over square and rectangular sections due to uniform stressing. In this test, specimen is placed between the bottom cross head and table of machine while in tensile test specimen is placed between the two cross heads. Procedure for performing the test is same as that for tensile test. On application of load at regular intervals, strain is measured at various points and stress strain curve is plotted (Fig. 1.10). Compression test is performed for specimen with limited length to avoid bending due to column section.

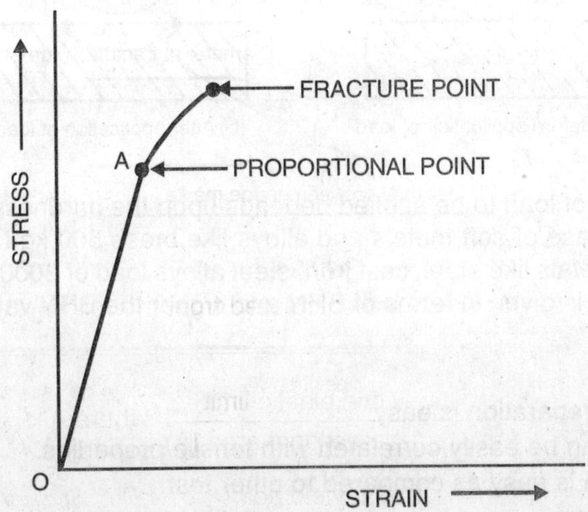

Fig. 1.10 Stress strain diagram for brittle material

(iii) **Hardness testing :** Hardness is the ability of a material to resist abrasion, indentation, wear and scratch. Most commonly used test to determine the hardness of materials are

1. Brinell hardness test
2. Rockwell hardness test
3 Vicker hardness test

1. **Brinell Hardness test :** In this test a hardened steel ball indenter of 10 mm diameter is pressed against a surface of test specimen by gradual application of load. The load is applied mechanically or hydraulically and maintained for 10–15 seconds. Ball make an impression on test specimen which is measured by microscope. Brinell Hardness Number (BHN) is calculated by applied load per unit surface area of indentation.

$$\text{BHN} = \frac{\text{Applied load}}{\text{Area of ball impression}}$$

$$\text{BHN} = \frac{2P}{\pi D\left(D - \sqrt{D^2 - d^2}\right)}$$

where P = Applied load in kg f
D = Diameter of ball in mm.
d = diameter of impression in mm
t = depth of impression

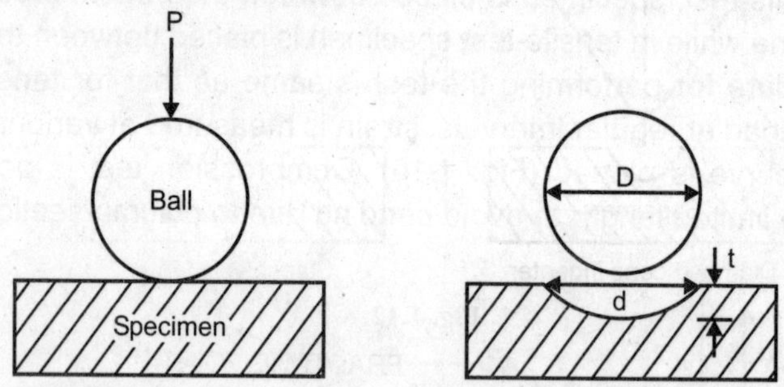

(a) Before application of load (b) After application of load

Fig. 1.11

The amount of load to be applied depends upon the hardness of test metal for example in case of soft metals and alloys like brass 500 kg load is applied while for hard metals like steel, cast iron, steel alloys load of 3000 kg is applied. Brinell hardness is given in terms of BHN and higher the BHN value harder will be the material.

Advantages

1. Surface preparation is easy.
2. Results can be easily correlated with tensile properties.
3. Procedure is easy as compared to other test.

Disadvantages

1. Time consuming
2. Heavy test equipment

3. Due to large size of indentation produced its nature is destructive.

4. Test is not reliable for very hard materials.

2. **Rockwell Hardness test :** In this test the size of indenter and applied load are much smaller than Brinell hardness test. Rockwell hardness test is fast and machine is also easy to operate so it is most widely used method.

The dial of machines have nine scale from A to H, and K. But B scale and C scale are widely used scale. B scale uses a steel ball indenter and used for testing the hardness of soft materials for eg. low and medium carbon steels, aluminum, grey cast iron etc. Its range is 1 to 100. C scale uses a diamond cone indenter and used for testing the hardness of hard materials for e.g., alloy cast irons, hardened steel etc.

Rockwell Hardness testing comprises of two stages. In first stage specimen is placed against steel ball indenter and initial load of 10 kg (called minor load) is applied, to help seat the specimen. Minor load causes small penetration into the surface. In second stage dial indicator is set at zero and major load of 60, 100 or 150 kg is applied to force the indenter into the specimen. When major load is removed dial still shows zero which indicates that minor load is still applied. Rockwell Hardness Number is measured from test gauge which corresponds to depth of penetration due to major load.

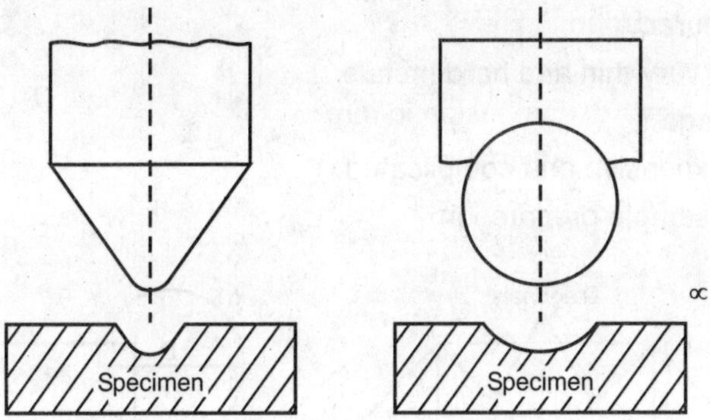

Diamond cone indenter Steel ball indenter

Fig. 1.12

Advantages

1. Less time consuming in comparision to Brinell Hardness test.

2. Impressions is small in size so nature is nondestructive.

3. Simple to operate and no requirement of microscope for diameter measurement.

4. Polishing of specimen is not required.

5. Free from personal errors.

6. More flexible than Brinell and can be used for testing of wide range of material.

Disadvantage

1. Surface preparation is typical

3. **Vicker's Hardness Test :** This method of hardness measurement is similar to Brinell hardness test but it eliminates the Brinell hardness test limitations of flatenning of steel ball indenters during the testing of harder materials.

In Vicker's Hardness test a square based diamond pyramid indenter with 136° angle between opposite faces is used.

The impression made by indenter on surface has square boundary with depression inside. The results obtained are more accurate because diagonal of square can be measured more accurately than the diameter of circle. Following relation is used to determine Vicker's hardness number (VHN)

$$VHN = \frac{\text{Load Applied}}{\text{Area of Impression}}$$

$$= \frac{2P \sin \theta/2}{d^2} = 1.8544 \cdot \frac{P}{d^2}$$

where P = Applied load

d = Average of two diagonals

θ = Angle between opposite faces of diamond pyramid = 136°

Advantages

1. High Accuracy.
2. Used for very thin and hard metals.

Disadvantage

1. Test is expensive and complicated.
2. Require sample preparation.

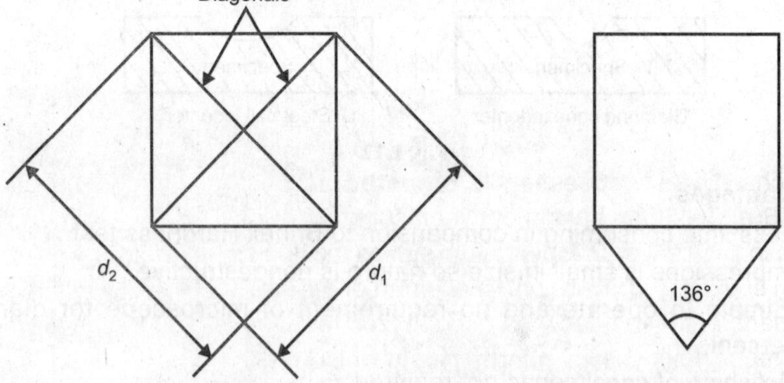

(a) Bottom view (b) Front view

Fig. 1.13

(iv) **Impact Test :** Impact tests are performed to study the behaviour of the material under high velocity suddenly applied load, also called as impact load. It measures the fracture energy of the material and indicates the relative toughness of the material.

Impact testing is performed on a pendulum type impact testing machine. First of all a notched specimen is prepared whose dimensions depends upon the type of test to be performed. The prepared specimen is placed on the anvils or supports of the machine and then striked by a heavy weight attached at the end of the pendulum. The pendulum falls from a standard height, swings along its arc, strikes and breaks the specimen and then rises to a certain height on the other side. By knowing the initial height and final height of specimen, the energy consumed in breaking the specimen can be calculated or can be directly determined from the calibrated scale of the machine.

Two types of test are generally performed on the machine.

1. Izod test 2. Charpy test

In Izod test initially a specimen is prepared as per the dimensions given in Fig. 1.14 (*a*). The specimen is then held in the supports or anvils in cantilever position as shown in Fig. 1.14 (*b*) and striked by a heavy mass attached at the bottom of the pendulum.

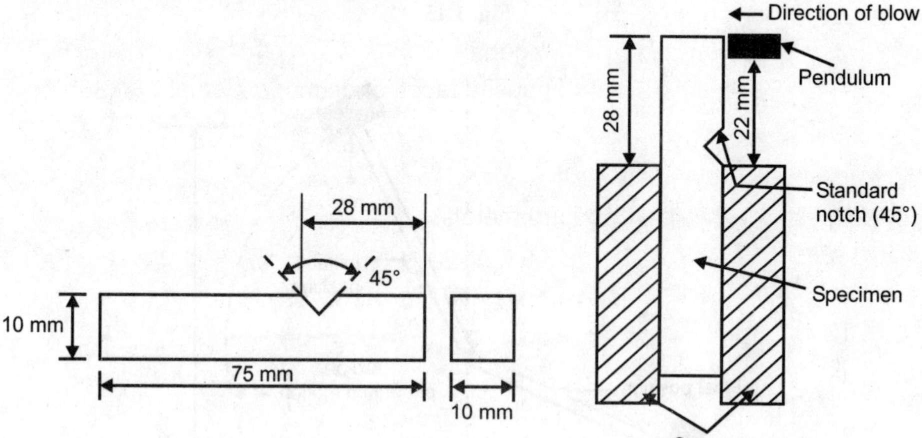

(a) Dimensions of Izod test specimen

(b) Cantilever beam position of specimen in izod test

Fig. 1.14

In charpy test, specimen is prepared as per the dimensions given in Fig. 1.15(*a*). It is then held in between supports or anvils in a simply supported position in such a way that pendulum strikes the specimen in a direction opposite to notch section [Fig. 1.15 (*b*)]. The length of the specimen in this test is smaller in comparison to the specimen used in Izod test. Schematic diagram of impact testing machine is shown in Fig. 1.16, in which

W = Weight of pendulum Hammer

h_1 = Initial height

h_2 = Final height

Now, energy consumed in fracture = difference of potential energy at height h_1 and h_2, so

$$\Delta E = E_1 - E_2$$

or

$$\Delta E = W(h_1 - h_2)$$

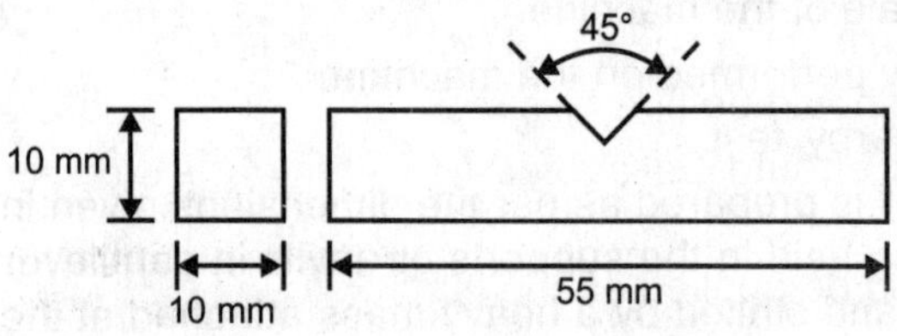

(a) Dimensions of Charpy test specimen

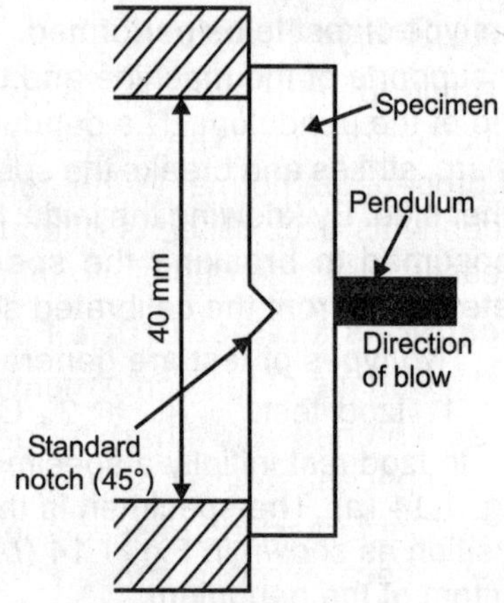

(b) Position of specimen in Charpy test

Fig. 1.15

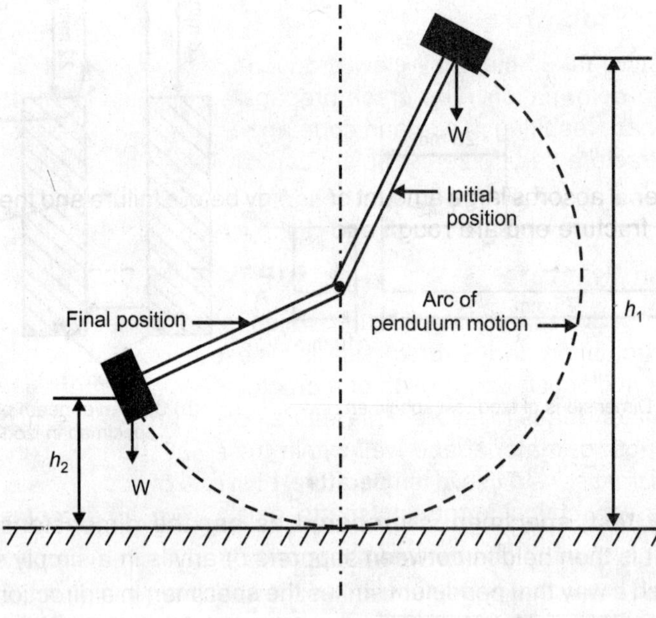

Fig. 1.16

1.8. FRACTURE

Fracture is defined as the separation of a solid into two or more parts under the action of an applied force. Fracture results as a sudden breaking of a material due to excessive plastic deformation. Fracture can be classified as

1. Brittle Fracture
2. Ductile Fracture
3. Fatigue Fracture
4. Creep Fracture

1.8.1. Brittle Fracture

Brittle fracture takes place by rapid propagation of a crack with little or no plastic deformation. The fracture crack starts from the location of strain raisers and propagates normal to the applied tensile stress. As the crack propagates suddenly, its occurrence is often unpredictable. Occurrence of brittle fracture increases with high rate of straining and decrease in temperature.

Precautions against Brittle Failure

1. Surface of the components should be fine finished.
2. Notches and sharp edges should be avoided to eliminate stress concentration.
3. Compressive stresses should be applied on the surface to mitigate the effect of tensile stress.
4. Stresses should be applied parallel to the direction of elongation.
5. Brittle fracture can also be reduced by tempering, chemical strengthening, etc.

1.8.2. Ductile Fracture

Ductile fracture takes place by slow propagation of crack after prolonged and appreciable deformation. The crack propagates in a direction 45° to the axis of tensile load, resulting in cup and cone appearance of fractured surface. Cup and cone fracture can be seen in plain carbon steel, mild steel, etc.

Ductile material absorbs large amount of energy before failure and the surface obtained at the fracture end are rough and dull.

1.8.3. Fatigue Fracture [UPTU Dec. 2008, May 2009]

Fatigue failure is defined as a fracture which takes place due to repeated application of fluctuating and reversing loads. These fluctuating and reversing loads promotes nucleation and growth of a crack, which becomes microscopic in size and ultimately leads to failure. This failure occurs at stresses much below the tensile strength of material and well within the elastic range. Tendency of fatigue failure increases with rise in temperature, high rate of straining etc. Fatigue consideration is very critical in rotating shafts, axles, wings of aeroplanes, etc.

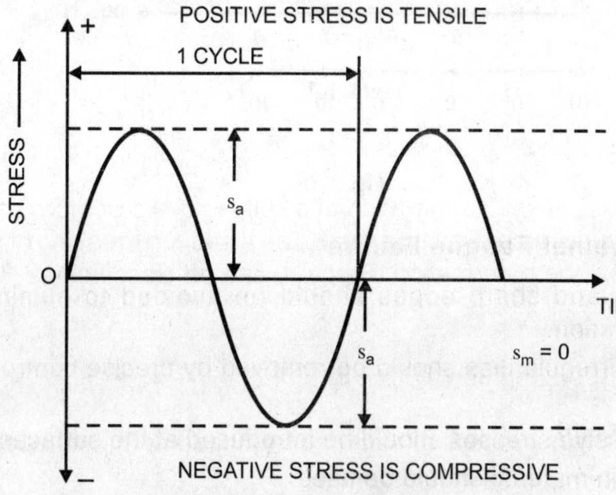

Fig. 1.17. Reversed stress (mean stress is zero).

The fatigue or endurance limit for a material is the maximum value of completely reversing stress that the standard specimen can withstand for an unlimited number of cycles without fatigue failure. The endurance limit is considered as a criterion of failure for fluctuating stresses.

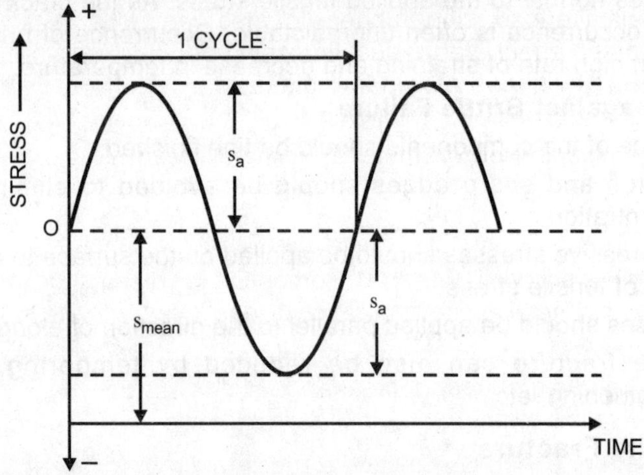

Fig. 1.18. Fluctuating stress.

The graph plotted between maximum applies stress (S) and the number of stress cycle (N) before fatigue failure on a log-log paper is known as S-N curve.

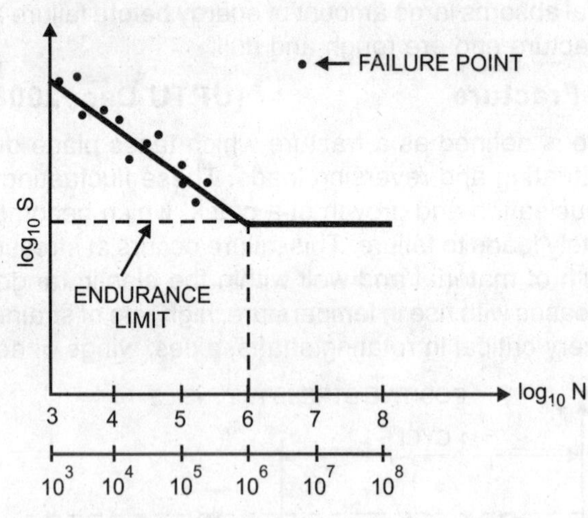

Fig. 1.19.

Precautions against Fatigue Failure

1. Notches and sharp edges should be avoided to eliminate stress concentration.
2. Surface irregularities should be removed by precise control of surface finish.
3. Compressive stresses should be introduced at the surfaces.
4. Fine grain material should be used.

1.8.4. Creep Fracture

Creep is a slow permanent deformation of a metal due to steady load or prolonged loadings. It is influenced by temperature so the materials like copper, iron and its alloys are susceptible to creep when subjected to high temperature. Some materials like lead, zinc, rubber etc. also creeps at room temperature. Examples of creep effect are gradual loosening of bolts, sagging of long span cables, greater thickness of window glass at bottom and deformation of components of machines over a long time.

Excessive creeping of the material ultimately leads to fracture known as creep fracture.

Creep Curve

Creep curve is shown in Fig. 1.20. Portion OA represents the instantaneous strain developed in metal due to application of load. Creep occurs in three stages:

(i) **Primary Creep :** In this stage, material initially strains rapidly but straining rate gradually decreases due to work hardening. Primary creep occurs at low temperature so it is also known as cold creep.

(ii) **Secondary Creep :** In second stage, strain rate is constant. Secondary creep occurs at high temperature so it is also known as hot creep.

(iii) **Tertiary Creep :** In third stage, material strains rapidly until it fractures.

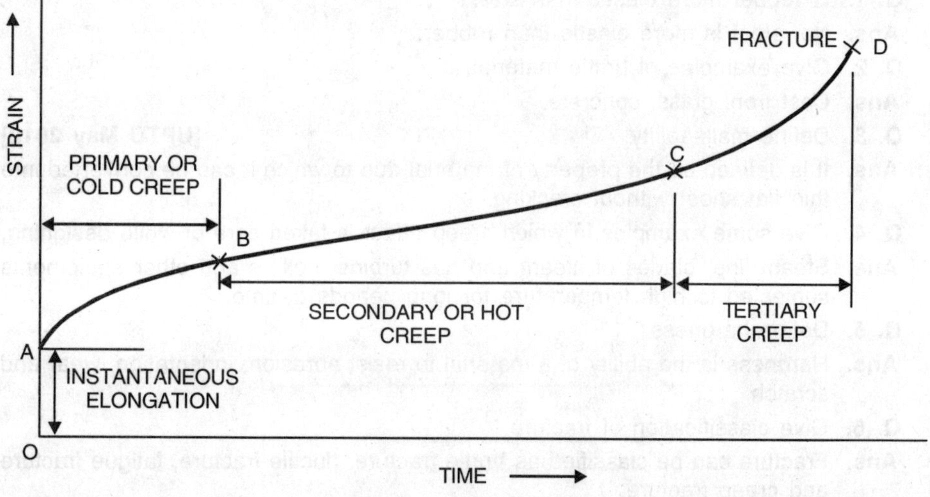

Fig. 1.20. Creep Curve.

SUMMARY

- Materials are classified into metals and non-metals.
- Metals are classified as ferrous metal and alloys, and non-ferrous metal and alloys.
- Nonmetals are classified as ceramics and organic polymers.
- Ferrous metal contain iron as their main constituent whereas non-ferrous metal contain metal other than iron as their main constituent.

- Resistance offered by a metal to external load is strength.
- Elasticity is the property due to which material regains its original shape and size.
- Stiffness resists deflection or deformation of materials.
- Due to ductility material can be drawn into wires.
- Due to malleability metal can be converted into thin sheets without cracking or failure.
- Brittleness shows the lack of plasticity.
- Toughness refers to strain energy absorbing capacity of material before failure.
- Hardness is the ability of a material to resist abrasion, indentation, wear, scratching, etc.
- Fracture is the separation of a metal into two or more parts.
- Fatigue failure occurs due to reversing and fluctuating loads.
- Creep is the permanent deformation of metal due to steady loads or prolonged loadings.

SHORT QUESTIONS WITH ANSWERS

Q. 1. Is rubber more elastic than steel?

Ans. No, steel is more elastic than rubber.

Q. 2. Give examples of brittle material.

Ans. Cast iron, glass, concrete.

Q. 3. Define malleability. **[UPTU May 2010]**

Ans. It is defined as the property of material due to which it can be converted into thin flat sheet without cracking.

Q. 4. Give some examples in which creep effect is taken care of while designing.

Ans. Steam line, blades of steam and gas turbine, boilers and other equipments subjected to high temperature for long periods of time.

Q. 5. Define hardness.

Ans. Hardness is the ability of a material to resist abrasion, indentation, wear and scratch.

Q. 6. Give classification of fracture.

Ans. Fracture can be classified as brittle fracture, ductile fracture, fatigue fracture and creep fracture.

Q. 7. Define metals.

Ans. Metals are generally solids and have good heat and electric conductivity, good malleability and ductility, high melting point and high strength.

Q. 8. Give classification of properties of materials.

Ans. (a) Physical properties

(b) Electrical properties

(c) Chemical properties

(d) Thermal properties

(e) Mechanical properties.

Q. 9. What is indicated by resilience?

Ans. Resilience indicates the capacity of material to withstand shock and vibrations.

Q. 10. Give examples of ductile material.

Ans. Copper, aluminium, mild steel.

Q.11. What is the main effect of fatigue on materials?

Ans. Loss of ductility and strength.

Q.12. List some applications in which effect of fatigue behaviour is critical?

Ans. Wings of aeroplanes, connecting rod of I.C. engines and leaf springs of automobiles.

Q. 13. Write any three types of Destructive test

Ans. (i) Tensile test (ii) Hardness test
(iii) Impact

Q. 14. What is the basic difference in Izod and Charpy test with respect to position of specimen.

Ans. In Izod test specimen is held in supports in cantilever position but in Charpy test specimen is held in between supports in simply supported position.

Q. 15. What type of indenter is used in Vicker's Hardness test.

Ans. In Vicker's Hardness test a square based diamond pyramid indenter with 136° angle between opposite faces is used.

Q. 16. What is the use of C scale in Rockwell Hardness test.

Ans. C scale is used for testing the hardness of hard materials e.g., alloy cast iron hardened steel etc.

Q. 17. What type of indenters are used by B scale and C scale in Rockwell hardness test.

Ans. B scale uses steel ball indenter and C scale uses diamond cone indenter.

Q. 18. Write the major difference between toughness and hardness of material.

Ans. Toughness refers to strain energy absorbing capacity of material before failure whereas hardness is the ability of material to resist abrasion indentation, wear, scratching.

REVIEW QUESTIONS

1. What do you mean by materials?
2. Give the brief classification of materials?
3. What do you mean by mechanical properties of materials? Explain them in brief.
4. Write short notes on :
   (a) Fracture (b) Creep
   (c) Fatigue
5. Compare the following : **[UPTU May 2012]**
   (a) Brittleness and Hardness (b) Strength and Stiffness
   (c) Ductility and Malleability (d) Hardness and Toughness
6. Explain the following terms :
   (a) Ductility (b) Stiffness
   (c) Brittleness (d) Fatigue
   (e) Malleability (f) Toughness
   (g) Resilience **[UPTU Dec. 2008]**

7. Draw rough sketch of stress strain diagrams of
   (a) Ductile material
   (b) Brittle material

   Briefly explain what do you understand by toughness.

   [UPTU Dec. 2008; May 2010]

8. Explain the following :
   (a) Ductile fracture (b) Brittle fracture
   (c) Creep fracture (d) Fatigue fracture

   [UPTU Dec. 2010]

9. What do you understand by creep. Discuss creep curve

   [UPTU May 2012]

SELF TEST QUESTIONS

A. State Whether True or False

1. Secondary creep is known as cold creep.
2. Steel is a nonferrous material.
3. Higher the strength the higher is the capacity of the material to withstand load without failure.
4. Ceramics are compounds of metallic and nonmetallic elements.
5. Stiffness is also known as rigidity.

B. Fill in The Blanks

1. The ability of material to resist abrasion, wear indentation is known as ....
2. ....is a property which resists deflection or deformation.
3. Ability of a material to be converted into thin flat sheets is known as..........
4. Resistance offered by a material to external load is known as ....
5. .... fracture takes place by rapid propagation of a crack with no plastic deformation.
6. .... is the capacity of a material to absorb energy in the elastic range.
7. .... is a slow permanent deformation of metal due to prolonged loadings.
8. Fatigue fracture takes place due to .... and .... loads.
9. Metals can be broadly classified as .... and ....
10. Primary creep occurs at .... temperature.
11. Cup and cone fracture occurs in ....
12. .... and .... should be avoided to eliminate stress concentration.
13. The ductility of a piece of a chalk is .... **[UPTU Dec. 2008]**
14. Mild steel is .... ductile than cast iron.
15. Energy required to rupture the material is known as ....

    [UPTU Dec. 2009]

C. Multiple Choice Questions

1. Fatigue failure occurs when a part is subjected to
   (a) Fluctuating stress (b) Compressive stress
   (c) Tensile stress (d) Uniform stress
2. Property of a material which recovers original shape and size after removal of load
   (a) Plasticity (b) Elasticity
   (c) Toughness (d) Ductility
3. Hardness of a material is tested by
   (a) Brinell test (b) Rockwell test
   (c) Both (a) and (b) (d) None
4. Which one is ceramic material ?
   (a) Iron (b) Aluminium
   (c) Steel (d) Cement.
5. Organic polymer is
   (a) Aluminium (b) Zinc
   (c) Copper (d) Rubber
6. Chemical properties are
   (a) Atomic weight (b) Shape
   (c) Size (d) Colour
7. Ferrous material is
   (a) Aluminium (b) Zinc
   (c) Steel (d) Copper
8. In malleability, plastic deformation is due to
   (a) Tensile load (b) Compressive load.
   (c) Both (a) and (b) (d) None
9. Deformation in ductility is due to
   (a) Tensile load (b) Compressive load
   (c) Both (a) and (b) (d) None
10. Degree of stiffness of material is represented by
    (a) Modulus of elasticity (b) Plasticity
    (c) Hooke's law (d) None
11. Brittleness shows the lack of
    (a) Elasticity (b) Plasticity
    (c) Resilience (d) Toughness
12. Strength is measured in terms of
    (a) Stress (b) Strain
    (c) Elasticity (d) Plasticity
13. Main constituent of ferrous metal is
    (a) Silver (b) Copper
    (c) Iron (d) Aluminium
14. Nonferrous metal is
    (a) Pig iron (b) Aluminium
    (c) Steel (d) Cast iron

15. Nonmetals which are basic oxides and are compounds of metallic and nonmetallic elements are known as
(a) Organic polymers (b) Ceramics
(c) Pig iron (d) Polyethylene

16. Materials having high molecular weight and complex three dimensional structure are known as
(a) Refractories (b) Concrete
(c) Ceramics (d) Organic polymer

17. Property which give the response of material under externally applied forces is
(a) Thermal (b) Mechanical
(c) Physical (d) Chemical

18. The ability of a material to offer resistance to externally applied load is
(a) Elasticity (b) Ductility
(c) Strength (d) Brittleness

19. Property of a material which enables permanent deformation of material without failure or rupture
(a) Elasticity (b) Ductility
(c) Malleability (d) Plasticity

20. The ability of material to absorb energy within elastic limit and withstand shock and vibrations is
(a) Resilience (b) Hardness
(c) Stiffness (d) Brittleness

21. Property of a material due to which it can be drawn into wires without rupture
(a) Plasticity (b) Malleability
(c) Brittleness (d) Ductility

22. Ability of a material to resist abrasion, indentation, wear and scratch
(a) Hardness (b) Stiffness
(c) Toughness (d) Resilience

23. Copper is
(a) Brittle material (b) Ductile material
(c) Ferrous material (d) None of these

24. Property of material due to which it can be converted into thin flat sheet without cracking
(a) Ductility (b) Stiffness
(c) Malleability (d) Plasticity

25. The property of material which refers to the strain energy absorbing capacity of material before its facture
(a) Hardness (b) Ductility
(c) Toughness (d) Plasticity

26. The property of material by virtue of which it fails or fractures without any appreciable deformation
(a) Ductility (b) Hardness
(c) Malleability (d) Brittleness

27. Which test is used to determine the hardness of material

(a) Brinell (b) Rockwell

(c) Vickers (d) All of these

28. The property of material due to which it resists deflection or deformation

(a) Plasticity (b) Stiffness

(c) Brittleness (d) None of these

29. Brittle material is

(a) Concrete (b) Cast iron

(c) Glass (d) All of these

30. Separation of solid into two or more parts under the action of applied force is called

(a) Hardness (b) Stiffness

(c) Fracture (d) Plasticity

31. With rise in temeprature, toughness of material

(a) Increases (b) Decreases

(c) Constant (d) None of these

32. The limit upto which stress is proportional to strain

(a) Elastic limit (b) Yield point

(c) Proportional limit (d) None of these

33. The fracture which take place by rapid propagation of a crack with little or no plastic deformation

(a) Brittle fracture (b) Ductile fracture

(c) Fatigue fracture (d) Creep fracture

34. Cup and cone fracture occurs in

(a) Glass (b) Cast iron

(c) Mild steel (d) Plastic

35. Fracture which takes place due to slow propagation of crack after prolonged and appreciable deformation.

(a) Brittle fracture (b) Ductile fracture

(c) Fatigue fracture (d) Creep fracture

36. The fracture which take place due to repeated application of fluctuating and reversing loads is

(a) Fatigue fracture (b) Ductile fracture

(c) Brittle fracture (d) Creep fracture

37. Maximum value of completely reversing stress that the standard specimen can withstand for an unlimited number of cycles without fatigue fracture is known as

(a) Elastic limit (b) Proportional limit

(c) Endurance limit (d) None of these

38. The graph between maximum applied stress and number of stress cycle before fatigue failure on log-log paper is

(a) Creep curve (b) SN curve

(c) Stress strain curve (d) None of these

39. The material in which creep occurs at high temperature

(a) Zinc (b) Rubber

(c) Lead (d) Copper

40. Slow permanent deformation of a metal due to steady load or prolonged loading leads to

(a) Ductile fracture (b) Fatigue fracture

(c) Creep fracture (d) Brittle fracture

41. Secondary creep occurs at

(a) High temperature (b) Low temperature

(c) Room temperature (d) None of these

42. In creep curve, strain rate is constant during

(a) Primary creep (b) Secondary creep

(c) Tertiary creep (d) None of these

43. Brittle failure occurs when a part is subjected to

(a) Compressive stress (b) Tensile stress

(c) Fluctuating stress (d) Uniform stress **[UPTU Dec. 2009]**

44. Ductile fracture occur when a part is subjected to

(a) Compressive stress (b) Tensile stress

(c) Fluctuating stress (d) Uniform stress **[UPTU May 2010]**

D. Match the Following

1.

(A)	(B)
(i) Resistance to penetration	(a) brittleness
(ii) Tendency to break under an impact of load	(b) ductility
(iii) Ability to be stretched before fracture	(c) stiffness
(iv) Ability of a material to resist deformation	(d) hardness

2.

(A)	(B)
(i) Strain energy absorbing capacity before failure	(a) Elasticity
(ii) Slow stretching or continuous plastic extension under steady stress	(b) Resilience
(iii) Property of material to return to its original size and shape on removal of load	(c) Toughness
(iv) Property of material to sustain shock load without permanent deformation.	(d) Creep

3.

(A)	(B)
(i) Ferrous metal	(a) PVC
(ii) Non-ferrous metal	(b) Cast iron
(iii) Ceramic	(c) Silicon carbides
(iv) Organic polymers	(d) Copper

4.

(A)	(B)
(i) Cast iron	(a) Proportional limit
(ii) Mild steel	(b) Degree of stiffness
(iii) Hooke's law	(c) Ductile material
(iv) Modulus of elasticity	(d) Brittle material

ANSWERS

A. True or False

1. False 2. False 3. True 4. True 5. True

B. Fill in the Blanks :

1. Hardness
2. Stiffness
3. Malleability
4. Strength
5. Brittle
6. Resilience
7. Creep
8. Fluctuating, reversing
9. Ferrous, nonferrous
10. Low
11. Plain carbon steel, mild steel.
12. Notches, Sharp corners
13. Very less or negligible
14. More
15. Impact toughness

C. Multiple Choice Questions

1. (a)	**2.** (b)	**3.** (c)	**4.** (d)	**5.** (d)
6. (a)	**7.** (c)	**8.** (b)	**9.** (a)	**10.** (a)
11. (b)	**12.** (a)	**13.** (c)	**14.** (b)	**15.** (b)
16. (d)	**17.** (b)	**18.** (c)	**19.** (d)	**20.** (a)
21. (d)	**22.** (a)	**23.** (b)	**24.** (c)	**25.** (c)
26. (d)	**27.** (d)	**28.** (b)	**29.** (d)	**30.** (c)
31. (b)	**32.** (c)	**33.** (a)	**34.** (c)	**35.** (b)
36. (a)	**37.** (c)	**38.** (b)	**39.** (d)	**40.** (c)
41. (a)	**42.** (b)			

D. Match the Following

1. (i)—(d), (ii)—(a), (iii)—(b), (iv)—(c) **2.** (i)—(c), (ii)—(d), (iii)—(a), (iv)—(b)

3. (i)—(b), (ii)—(d), (iii)—(c), (iv)—(a) **4.** (i)—(d), (ii)—(c), (iii)—(a), (iv)—(b)

❑❑❑

UNIT-II

Basic Metals and Alloys: Properties and Applications

2. FERROUS MATERIALS

3. HEAT TREATMENT

4. NON-FERROUS MATERIALS

SYLLABUS

Ferrous Materials : Carbon Steels, Its Classification Based on % Carbon as Low, Mild, Medium and High Carbon Steel, its Properties and Applications. Wrought Iron. Cast Iron. Alloy Steels: Stainless Steel, Tool Steel. Elementary Introduction to Heat-Treatment of Carbon Steels: Annealing, Normalizing, Quenching and Tempering and Casehardening.

Non-Ferrous Metals and Alloys: Common Uses of Various Non-ferrous Metals and Alloys and its Composition such as Cu-alloys: Brass, Bronze, Al-alloys such as Duralumin.

CHAPTER **2**

Ferrous Materials

INSIDE THIS CHAPTER

2.1. INTRODUCTION

Ferrous materials are those which contain iron as their main constituent and are most widely used because of their low cost and varied mechanical properties. Other constituents such as carbon, manganese, silicon, sulphur and phosphorous exist in varying proportions with iron to form various ferrous materials such as pig iron, wrought iron, cast iron, alloy steel, carbon steel, etc.

In this chapter we shall discuss about wrought iron, cast iron and steel.

2.2. WROUGHT IRON

Pig iron is the crude impure iron which is extracted from iron ores by melting it in a blast furnace. Pig iron is the first usable form of iron and is used to produce the wrought iron by remelting it in reverberatory furnace also called as puddling furnace.

The furnace is made up of refractory material and lined inside with haematite (Fe_2O_3). The roof of the furnace is peculiar in shape to increase the contact of the flame with hearth. Pig iron piled on the hearth is melted by the gas flame and is collected from the furnace in a spongy form known as puddle ball. Puddle ball are further worked to remove the slag and impart the desired properties.

Iron thus obtained is known as wrought iron and contains 99.8% of pure iron.

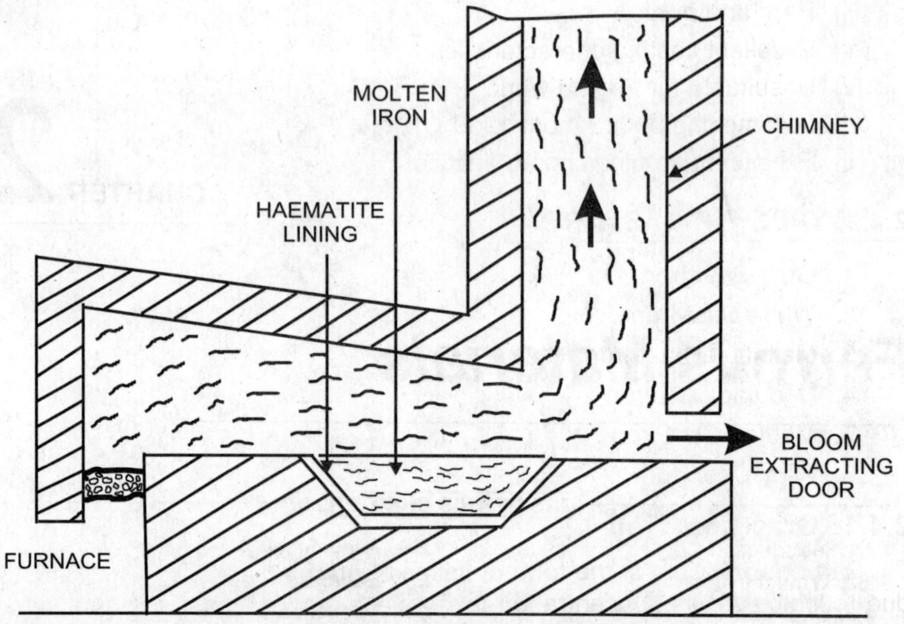

Fig. 2.1. Puddling or Reverberatory Furnace

2.2.1. Properties

(i) It has excellent corrosion resistance.
(ii) It has high ductility and can be easily forged.
(iii) It has good machinablity and weldability.
(iv) Its melting point is about 1500°C.
(v) It can neither be hardened nor tempered.
(vi) It can never be cast.

2.2.2. Applications

(i) In forged articles.
(ii) In manufacturing of iron bars, forging blooms and billets.
(iii) In manufacturing of water and steam pipes.
(iv) In manufacturing of nuts, bolts, chain, etc.

2.3. CAST IRON

Cast iron is an alloy of iron and carbon having a minimum of 2% to maximum of 4.5% carbon with small amounts of silicon (1% to 3%), sulphur (upto 0.15%), phosphorus (upto 1%) and manganese (0.5% to 1%).

Cast iron is manufactured by remelting pig iron with a definite amount of limestone, coke, spoiled castings, etc. in cupola furnace.

2.3.1. Properties

(i) Low resistance to tension.
(ii) High compressive strength.

(iii) Hard and brittle.
(iv) Excellent casting characteristics.
(v) Unsuitable for forging work.
(vi) Low melting point (about 1200°C).
(vii) Excellent damping characteristics.

2.4. TYPES OF CAST IRON

1. Gray Cast Iron
2. White Cast Iron
3. Malleable Cast Iron
4. Nodular Cast Iron
5. Mottled Cast Iron
6. Alloy Cast Iron

2.4.1. Gray Cast Iron

Carbon is present in the form of flakes of graphite. Flakes have sharp tips due to which stress concentration takes place and makes it brittle in nature. It gives gray fracture due to presence of graphite.

Composition

Carbon 3 to 3.5%
Silicon 1 to 2.75%
Phosphorus 0.15 to 1%
Sulphur 0.02 to 0.15%
Manganese 0.4 to 1%
Rest percentage is of iron.

Properties

(i) Due to free graphite it has self lubricating properties.
(ii) Good machinability and fusibility.
(iii) Low tensile strength and good compressive strength.
(iv) Brittle in nature.
(v) Good resistance to wear.
(vi) Excellent damping capacity.

Uses of Gray Cast Iron

Machine tool structures, pipes, pipe fittings, manhole covers, piston, rings, cylinder head, flywheel, parts subjective to compressive stress such as base of machines, etc.

2.4.2. White Cast Iron

Due to the presence of carbon in the form of carbide it shows white colour at fracture. This carbide is known as cementite and is extremely hard and brittle. Cementite formation takes place due to rapid cooling of molten iron.

Composition

Carbon 1.75 to 2.3%, silicon 0.85 to 1.2%, manganese upto 0.5%, phosphorus 0.05 to 0.2%, sulphur upto 0.12% and remaining percentage is of iron.

Properties

(i) Extremely hard and brittle.
(ii) Poor machinablity.
(iii) High tensile and low compressive strength.
(iv) Low melting point than gray cast iron.

Uses

(i) In parts subjected to excessive wear or where hard coating is required e.g. outer surface of car wheel, wearing plate, cams, balls, brake block, etc.
(ii) For inferior casting.
(iii) As raw material for malleable cast iron and wrought iron.

2.4.3. Malleable Cast Iron

It is made from white cast iron by a heat treatment process known as annealing. Due to annealing cementite decomposes to ferrite and produces malleable cast iron.

Composition

Carbon 2 to 3%
Silicon 0.6 to 1.3%
Manganese 0.2 to 0.6%
Phosphorus upto 0.18%
Sulphur upto 0.18%
Rest percentage is of iron.

Properties

(i) Ductile and can be bent without breaking.
(ii) High yield strength.
(iii) Good wear resistance.
(iv) Easily machinable.

Uses

Agricultural machinery, crankshafts, gear housing, hubs of wagon wheels, pipe fittings, etc.

2.4.4. Nodular Cast Iron

Nodular cast iron is produced by converting the graphite of cast iron from flake form to nodular form by little addition of magnesium to molten gray cast iron. This iron is also known as spheroidal graphite cast iron or ductile cast iron.

Composition

Carbon 3.2 to 4.2%

Silicon 1 to 2.8%

Manganese 0.3 to 0.8%

Phosphorus 0.03 to 0.1%

Magnesium 0.05 to 0.1%

Rest percentage is of iron.

Properties

(i) Good machinablity.

(ii) Excellent castability due to high fluidity.

(iii) High ductility than gray cast iron.

(iv) Its damping capacity lies between cast iron and steel.

(v) Good wear and shock resistance

Uses

Cylinder heads, pumps and compressors, pipes and pipe fittings, punch and dies, etc.

2.4.5. Mottled Cast Iron

It is a mixture of gray and white cast iron. Composition, colour, strength and hardness of mottled cast iron depend upon the ratio of free carbon to combined carbon. In general, carbon contents are divided equally in free and combined state and shows grayish white colour fracture.

2.4.6. Alloy Cast Iron

To achieve the desired properties in cast iron different alloying elements like nickel, silicon, sulphur, manganese are added to produce alloy cast iron. Due to alloying the properties are greatly improved i.e. high strength, high corrosion resistance, improved wear resistance, etc.

Alloy cast iron find extensive application in automobile parts such as cylinders, pistons, piston rings, etc.

2.5. STEEL [UPTU DEC. 2008, MAY 2009]

Steel is an alloy of carbon and iron having carbon upto 1.5% in combined form of iron carbide. These steels are known as plain carbon steels because their properties are mainly due to carbon content. Steels can be classified on the following basis :

(a) Depending upon the percentage of carbon

1. **Dead mild steel :** Carbon is upto 0.15%
2. **Low carbon steel or mild steel :** Carbon ranges from 0.15 to 0.3%
3. **Medium carbon steel :** Carbon ranges from 0.3 to 0.6%
4. **High carbon steel :** Carbon ranges from 0.6 to 1.5%

(b) Depending upon eutectoid reaction

1. **Hypoeutectoid steel :** Percentage of carbon is less than 0.8%
2. **Eutectoid steel :** Percentage of carbon is 0.8%

3. **Hypereutectoid steel :** Percentage of carbon ranges from 0.8 to 1.5%.

(c) Depending upon contents

1. Plain carbon steel
2. Alloy steel.

2.5.1. Dead Mild Steel

If the carbon content in steel is upto 0.15% then it is known as dead mild steel. It is used for making high ductility wire, thin sheets, wire, rod, welded and solid drawn tubes, chains, stampings, etc.

2.5.2. Low Carbon or Mild Steel

If carbon content in steel ranges from 0.15 to 0.30% then it is known as low carbon or mild steel.

Properties

(i) Soft and offers good ductility.
(ii) Tough but have low wear resistance.
(iii) Excellent formability.
(iv) It is not hardened by heat treatment but in some exceptional cases carburizing and annealing can be done.
(v) Low cost than other carbon steel.

Application

It is used for making screws, nuts, bolts, rivets, thin canes, chains, small forgings, wires, stampings of automobiles, etc.

2.5.3. Medium Carbon Steel

It contains 0.3 to 0.6% of carbon and is stronger than mild steel.

Properties

(i) Harder but less ductile than mild steel.
(ii) More response to heat treatment as compared to mild steel.
(iii) Good tensile strength i.e., 600-710 N/mm^2 can be achieved by various heat treatment processes.
(iv) Can be welded or brazed.

Applications

Used for making stronger screws, nuts, bolts, rivets, shafts, connecting rods, gears, crankshafts, axles, drop forgings, etc.

2.5.4. High Carbon Steel

It contains 0.6 to 1.5% of carbon and give good response to various heat treatment processes.

Properties

1. Highest tensile strength and hardness in plain carbon steels.

2. Mechanical properties can be easily varied by various heat treatment process.
3. Lowest ductility a nd machinablity in plain carbon steels.

Applications

Due to good tensile strength and hardness, high carbon steel find extensive application in hand tools, cutting tools, drills, milling cutters, wood working tools, punches, die taps, wheels, springs etc.

2.6. ALLOY STEEL

Alloy steel is steel in which elements other than carbon such as nickel, chromium, manganese, molybdenum, tungsten, vanadium are added to obtain the desired properties. The alloying of steel is done to accomplish one or more of the following objectives :

1. To increase the strength and hardness.
2. To improve resistance to abrasion and wear.
3. To improve corrosion resistance.
4. To improve ductility and machinability.
5. To improve electrical and mechanical properties.

Effects of various alloying elements are as follows :

1. **Nickel :** Increases tensile strength and toughness, improves shock and fatigue resistance.
2. **Chromium :** Increases hardness and corrosion resistance, provides stainless property and reduces the unwanted effects of sulphur.
3. **Tungsten :** Improves cutting hardness and red hardness, imparts abrasion resistance.
4. **Vanadium :** Increases hardness, yield strength and tensile strength.
5. **Sulphur :** Improves machinability, increases brittleness.
6. **Silicon :** Increases strength, hardness and resistivity.
7. **Manganese :** Increases toughness, hardness, strength and reduce brittleness due to sulphur.
8. **Cobalt :** Improves heat resistance, maintains steel hard at high temperature, increases hardness and strength.
9. **Boron :** Increases hardenability.
10. **Copper :** Increases corrosion resistance and strength.
11. **Molybdenum :** It form carbides and increases strength and impact resistance particularly at high temperature.
12. **Aluminium :** Increases hardness by forming extremely hard nitrides, used as deoxidiser.
13. **Titanium :** Prevents precipitation of chromium carbide, used as deoxidiser.
14. **Niobium :** Improves ductility and impact strength, decreases hardenability.

2.7. STAINLESS STEEL

The steel which resist the oxidation and corrosion effect due to the presence of chromium with small quantity of nickel, manganese, etc. is known as stainless steel. Chromium present in steel forms a fine film of chromium oxide on the surface which reduce further oxidation, rust or corrosion, and makes steel stainless. Stainless steel can be classified as

1. Ferritic Stainless Steel
2. Martensitic Stainless Steel
3. Austenitic Stainless Steel

2.7.1. Ferritic Stainless Steel

It contains 11 to 27% chromium, 1 to 1.5% manganese, 1% silicon with 0.1 to 0.2% carbon. They have high resistance to corrosion, good weldability, low impact strength and ductility. These steels are used in dairy equipments, heat exchangers, kitchen utensils, surgical equipments etc.

2.7.2. Martensitic Stainless Steel

It contains 12 to 18% chromium, manganese upto 1%, silicon upto 0.4% with 0.1 to 0.8% carbon. It has low corrosion resistance than ferritic stainless steel but can be hardened by heat treatment process. It is used in springs, shafts, bolts, nuts, rivets, screws, valves, scissors, etc.

2.7.3. Austenitic Stainless Steel

It contains 16 to 26% chromium, 2% manganese, silicon upto 2%, 3.5% to 22% nickel with 0.03 to 0.25% carbon. It is used in pump set, chemical plants, and kitchen utensils due to high corrosion resistance.

2.8. TOOL STEELS

Tool and die steels are special alloy steels which have high hardness and impact toughness, greater abrasion and wear resistance, low coefficient of friction at room and elevated temperatures.

Selection of proper tool depends on operation to be performed, material to be cut, rate of cutting, etc. SAE has classified tool steel in following classes.

2.8.1. Water Hardening Tool Steels

It contains 0.7 to 1.5% carbon, 0.4 to 0.5% manganese. These are used for chisels, drills, files and small tools.

2.8.2. Shock Resistant Tool Steels

It contains 0.45 to 0.55 % carbon, 0.2 to 0.4% manganese, 1.25 to 1.75% chromium. These are used in pneumatic tools, punches, heating dies, etc.

2.8.3. Cold Work Tool Steels

It contains 0.85% to 0.95% carbon, 1 to 1.3% manganese, 0.4 to 0.6% chromium. These are used in twist drills, milling cutters, boring tools, gauges, dies, etc.

2.8.4. Hot Work Tool Steels

It contains 0.3% carbon, 0.3% chromium, 0.3% molybdenum and finds application in hot drawing, hot forging, casting dies, hot blanking dies etc.

2.9. HIGH SPEED STEEL

It is capable of cutting metal at a much higher speed than other steels so known as High Speed Steel (HSS). High speed steels maintains their hardness even when heated upto 650°C. It contains tungsten as the main alloying element along with small quantities of cobalt, vanadium, chromium, etc. Different types of HSS are :

1. 18-4-1 High Speed Steel

It contains 0.75% carbon, 18% tungsten, 4% chromium and 1% vanadium. It is all purpose tool steel and most widely used in tools of lathe, shaper, planer, milling and drilling machine.

2. Molybdenum High Speed Steel

It contains 6% tungsten, 6% molybdenum, 4% chromium and 2% vanadium. It is used in drilling and tapping operations.

3. Super High Speed Steel

It contains 20% tungsten, 4% chromium, 2% vanadium and 12% cobalt. It is particularly used when the tool is subjected to high temperature and pressure.

2.10. SPECIAL PURPOSE TOOL STEEL

It comprises a variety of alloys like chromium, nickel, tungsten, molybdenum. These are used for special purposes e.g. to produce stainless and heat resisting components.

SUMMARY

- Ferrous materials are those which contain iron as their main constituent.
- Pig iron is the crude impure iron which is extracted from iron ores by melting it in a blast furnace.
- Wrought iron is obtained by remelting pig iron in reverberatory or puddling furnace.
- Cast iron is manufactured by remelting pig iron with a definite amount of limestone, coke, spoiled castings, etc. in cupola furnace.
- In gray cast iron carbon is present in the form of flakes of graphite.
- In white cast iron carbon is present in the form of carbide known as cementite.
- Malleable cast iron is made by annealing of white cast iron.
- In nodular cast iron, graphite of cast iron is converted from flake form to nodular form.
- Mottled cast iron is a mixture of gray and white cast iron.

- Steel is an alloy of carbon and iron having carbon upto 1.5%.
- Alloy steel is steel in which elements other than carbon such as nickel, chromium, manganese, molybdenum, tungsten, vanadium are added to obtain the desired properties.
- The steel which resist the oxidation and corrosion effect due to the presence of chromium with small quantity of nickel, manganese, etc. is known as stainless steel.
- Tool and die steels are special alloy steels which have high hardness and impact toughness, greater abrasion and wear resistance, low coefficient of friction at room and elevated temperatures.
- High speed steel generally contains 0.75% carbon, 18% tungsten, 4% chromium and 1% vanadium.
- Special purpose tool steel comprises a variety of alloys like chromium, nickel, tungsten, molybdenum, etc. to produce steel for specific purpose.

SHORT QUESTIONS WITH ANSWERS

Q. 1. In which form carbon is present in gray cast iron?

Ans. Carbon is present in form of flakes of graphite in gray cast iron.

Q. 2. Which element is added to steel to increase machinability?

Ans. Sulphur is added to increase the machinablity of steel.

Q. 3. What is the percentage of carbon in cast iron?

Ans. 2 to 4.5%.

Q. 4. Why gray cast iron is brittle?

Ans. Carbon is present in form of flakes of graphite. Flakes have sharp tips due to which stress concentration takes place and makes gray cast iron brittle.

Q. 5. What is the purpose of adding molybdenum to steel?

Ans. Molybdenum is added to increase strength and impact resistance particularly at high temperature.

Q. 6. Which alloy element is added to increase hardness and corrosion resistance of steel?

Ans. Chromium is added to increase the hardness and corrosion resistance of steel.

Q. 7. How nodular cast iron is produced?

Ans. Nodular cast iron is produced by converting the graphite of cast iron from flake form to nodular form by little addition of magnesium to molten cast iron.

Q. 8. Differentiate between plain carbon steel and alloy steel.

Ans. Plain carbon steels are those steels which have their properties mainly due to carbon contents whereas properties of alloy steel are mainly due to elements other than carbon such as manganese, chromium, etc.

Q. 9. What is the difference between hypoeutectoid and hypereutectoid steels?

Ans. Hypoeutectoid steel contains less than 0.8% carbon whereas hypereutectoid steel contains more than 0.8% of carbon.

Q. 10. Why the roof of the puddling furnace is peculiar in shape?

Ans. Roof of puddling furnace is made peculiar in shape to increase the contact of flame with hearth.

REVIEW QUESTIONS

1. What is wrought iron? Explain its properties and applications. [UPTU Dec. 2009]
2. Explain different types of cast iron. [UPTU Dec. 2009]
3. What is steel? Give the classification of steel.
4. What is alloy steel? Discuss the effect of various alloying element on steel.
5. What is stainless steel? Explain its types with examples.
6. Write short notes on [UPTU Dec. 2009]
   (a) Tool steel
   (b) High speed steel
7. Based on % of carbon, classify carbon steels. How do stainless steel become stainless ? [UPTU Dec. 2008]
8. What are high speed steels. What are the principal alloying elements in them. Describe their properties in brief. [UPTU Dec. 2011]
9. State the differnece between steel and cast iron with respect to their composition, tensile strength, hardness and toughness. [UPTU Dec. 2011]

SELF TEST QUESTIONS

A. State Whether True or False

1. Wrought iron has better corrosion resistance than steel.
2. Cast iron is suitable for forging work.
3. Gray cast iron has self-lubricating property.
4. Maximum percentage of carbon in steel is 1.5%.
5. Mild steel is not hardened by heat treatment.
6. Cementite is soft and brittle.
7. The percentage of carbon in mild steel is 0.7%.
8. Pig iron is used to form wrought iron.
9. Nodular cast iron has low ductility than gray cast iron.
10. Gray cast iron is brittle in nature.
11. Castability of cast iron is poor. [UPTU May 2010]
12. Stainless steel is an alloy steel. [UPTU May 2010]
13. Machinability of nickel and steel is good.
14. The castablity of Al is very poor. [UPTU May 2011]

B. Fill in The Blanks

1. Puddling furnace is lined inside with .... [UPTU Dec. 2010]
2. Gray cast iron gives gray fracture due to ....
3. To produce parts subjected to excessive wear........... is used.
4. In high carbon steel carbon ranges from .... to ....
5. Alloying element responsible for providing stainless property is ....
6. High speed steel can cut metal at much .... rate than other steel.

7. Percentage of carbon in eutectoid steel is ....
8. Carbide present in white cast iron is known as ....
9. .... is the main constituent of plain carbon steel.
10. High speed steel has ....% of chromium.

C. Multiple Choice Questions

1. Wrought iron is produced by remelting pig iron in
   (a) Cupola (b) Puddling furnace
   (c) Bessemer Converter (d) None
2. Melting point of wrought iron is
   (a) About 1500°C (b) About 1800°C
   (c) About 800°C (d) About 1200°C
3. Cast iron has unique property of
   (a) Hardness (b) Ductility
   (c) Malleability (d) Damping
4. Cast iron is produced by remelting pig iron in
   (a) Cupola (b) Puddling furnace
   (c) Bessemer Converter (d) None
5. In gray cast iron, carbon is present in form of
   (a) Flakes of graphite (b) Carbide
   (c) Nodular aggregates (d) Spheroids
6. In white cast iron, carbon is present in form of
   (a) Flakes of graphite (b) Carbide
   (c) Nodular aggregates (d) Spheroids
7. In nodular cast iron, carbon is present in form of
   (a) Flakes of graphite (b) Carbide
   (c) Nodular aggregates (d) Spheroids
8. Mottled cast iron is a mixture of
   (a) Malleable and white cast iron
   (b) Gray and white cast iron
   (c) White and nodular cast iron
   (d) Alloy and nodular cast iron
9. Gray cast iron is used in machine tool structure due to
   (a) Brittleness (b) Good damping capacity
   (c) Good machinability (d) Low tensile strength
10. Most extensively used material in hand tools, cutting tools is
    (a) Low carbon steel (b) Medium carbon steel
    (c) High carbon steel (d) None
11. Percentage of carbon in hypoeutectoid steel is
    (a) less than 0.8% (b) 0.8 to 1.5%
    (c) equal to 0.8% (d) None
12. The process which is used to form malleable cast iron from white cast iron is
    (a) Normalizing (b) Annealing
    (c) Case hardening (d) Tempering

13. The % of carbon in medium carbon steel varies

(a) Less than 0.20 (b) 0.10 to 0.30

(c) 0.30 to 0.60 (d) More than 0.60 **[UPTU Dec. 2008]**

14. Stainless steel comes under the category of **[UPTU May 2011]**

(a) Nonferrous metal and alloy (b) Ferrous metal and alloy

(c) Polymer (d) Composite

15. Crude impure iron extracted from iron ores by melting it in a blast furnace is

(a) Cast iron (b) Pig iron

(c) Wrought iron (d) Grey cast iron

16. Refractory material used for inside lining of puddling furnace.

(a) Cryolite (b) Bauxite

(c) Haematite (d) Soda lime

17. Percentage of iron in wrought iron is

(a) 90% (b) 93.5%

(c) 80% (d) 99.8%

18. Percentage of carbon in cast iron is

(a) 5 to 10% (b) 2 to 4.5%

(c) 10 to 15% (d) 0.5 to 2%

19. Cementite is

(a) Soft and brittle (b) Hard and brittle

(c) Ductile and soft (d) Hard and ductile

20. Carbon is present in form of carbide in

(a) Grey cast iron (b) White cast iron

(c) Alloy cast iron (d) None of these

21. Which cast iron is formed by annealing of white cast iron

(a) Malleable cast iron (b) Gray cast iron

(c) Nodular cast iron (d) Mottled cast iron

22. Percentage of carbon in Gray cast iron is

(a) 1 to 2% (b) 2 to 3%

(c) 3 to 3.5% (d) 3.5 to 5%

23. Steel which contain 0.8% carbon is

(a) Eutectoid steel (b) Hypoeutectoid steel

(c) Hypereutectoid steel (d) Mild steel

24. Which alloying element prevents precipitation of chromium carbide

(a) Silicon (b) Boron

(c) Titanium (d) Nickel

25. Which alloying element improves cutting hardness and red hardness

(a) Tungsten (b) Niobium

(c) Copper (d) Nickel

26. The steel which resist the oxidation and corrosion effect due to the presence of chromium with small quantity of nickel, manganese, etc. is

(a) Mild steel (b) High carbon steel
(c) Stainless steel (d) Low carbon steel

27. Steel which is used for cutting the metal at a much higher speed is

(a) Stainless steel (b) High speed steel
(c) Low carbon steel (d) Medium carbon steel

28. What is the % of chromium in 18–4–1 high speed steel

(a) 1% (b) 6%
(c) 18% (d) 4%

D. Match the Following

1.

(A)	(B)
(i) Wrought iron	(a) Improve cutting hardness
(ii) Cast iron	(b) Puddling furnace
(iii) Titanium	(c) Cupola furnace
(iv) Tungsten	(d) Deoxidiser

2.

(A)	(B)
(i) Dead mild steel	(a) 1.75 to 2.3% of carbon
(ii) White cast iron	(b) 3 to 3.5% of carbon
(iii) Medium carbon steel	(c) Upto 0.15% carbon
(iv) Gray cast iron	(d) 0.30% to 0.60% of carbon

3.

(A)	(B)
(i) Gray cast iron	(a) Annealing of white cast iron
(ii) White cast iron	(b) Ductile cast iron
(iii) Nodular cast iron	(c) Flakes of graphite
(iv) Malleable cast iron	(d) Carbide

4.

(A)	(B)
(i) Hypoeutectoid steel	(a) % of carbon is 0.8
(ii) Eutectoid steel	(b) % of carbon is less than 0.8
(iii) Hypereutectoid steel	(c) % of carbon is from 0.6 to 1.5
(iv) High carbon steel	(d) % of carbon is from 0.8 to 1.5

5.

(A)	(B)
(i) Dead mild steel	(a) Drills, milling cutters
(ii) Low carbon steel	(b) Crank shafts, gears
(iii) Medium carbon steel	(c) Thin sheets, wire, rod
(iv) High carbon steel	(d) Nuts and bolts

6.

(A)	(B)
(i) Nickel	(b) Improves machinability
(ii) Chromium	(b) Used as deoxidiser
(iii) Sulphur	(c) Increases tensile strength
(iv) Aluminium	(d) Increases handness and corrosion resistance

ANSWERS

A. True or False

1. True	2. False	3. True	4. True	5. True
6. False	7. False	8. True	9. False	10. True
11. False	12. True	13. False	14. False	

B. Fill in the Blanks

1. Haematite	6. Higher
2. Graphite	7. 0.8%
3. White cast iron	8. Cementite
4. 0.6 to 1.5%	9. Carbon
5. Chromium	10. 4%

C. Multiple Choice Questions

1. (b)	2. (a)	3. (d)	4. (a)	5. (a)
6. (b)	7. (c)	8. (b)	9. (b)	10. (c)
11. (a)	12. (b)	13. (c)	14. (b)	15. (b)
16. (c)	17. (d)	18. (b)	19. (b)	20. (b)
21. (a)	22. (c)	23. (a)	24. (c)	25. (a)
26. (c)	27. (b)	28. (d)		

D. Match the Following

1. (i)—(b), (ii)—(c), (iii)—(d), (iv)—(a) 2. (i)—(c), (ii)—(a), (iii)—(d), (iv)—(b)
3. (i)—(c), (ii)—(d), (iii)—(b), (iv)—(a) 4. (i)—(b), (ii)—(a), (iii)—(d), (iv)—(c)
5. (i)—(c), (ii)—(d), (iii)—(b), (iv)—(a) 6. (i)—(c), (ii)—(d), (iii)—(a), (iv)—(b)

❑❑❑

CHAPTER 3

Heat Treatment

INSIDE THIS CHAPTER

3.1. INTRODUCTION

Heat treatment is an operation or combination of operations involving heating and cooling of metals or alloys in its solid state in order to obtain certain physical and mechanical properties. Heat treatment is performed to obtain specific properties by changing shape, size, nature and distribution of different constituent in microstructure of metals.

3.2. OBJECTIVES OF HEAT TREATMENT [UPTU MAY 2009]

1. To relieve internal stresses set up during welding, casting, cold working, etc.
2. To change internal structure to improve their resistance to wear, heat, corrosion, etc.
3. To improve mechanical properties like toughness, strength, etc.
4. To drive out trapped gases from the material.
5. To modify various electrical and magnetic properties.
6. To soften metals for further treatment such as wire drawing and cold rolling.
7. To remove segregation.
8. To change and refine grain size.

3.3. PROCESS OF HEAT TREATMENT

Heat treatment comprises of following three steps.

1. Heating of the metal to specified temperature.

2. Holding the metal at high temperature for a specified period.
3. Cooling the metal at specified rate in a suitable medium.

Step 1

Metal or alloy is heated to specified temperature depending upon its grain size, grade, type and shape. This heating results in plastic deformation of grains and the extent of plastic deformation depends upon chemical composition of metal and temperature upto which it is heated.

Step 2

Hold the metal at increased temperature for a specified period so that structure of metal becomes uniform throughout the mass. Holding time period depends upon shape and size of component.

Step 3

It is the most important stage of heat treatment in which transformation in the properties of component takes place by cooling of metal. This transformation depends upon cooling rate and cooling medium. Generally water, oil, air, brine, caustic soda solutions are used as cooling medium.

3.4. MICRO CONSTITUENTS OF STEEL

1. α Ferrite

α ferrite is a solid solution of carbon in α iron. It has body centered cubic (BCC) structure. Carbon is slightly soluble in ferrite. The solubility of carbon in ferrite is 0.025 percent at 723°C. It is very soft, ductile and highly magnetic. It can not be hardened by rapid cooling.

2. Austenite

Austenite is a solid solution of carbon in gamma (γ) iron and has face centered cubic (FCC) structure. It has much greater solid solubility (max. 2% at 1130°C) for carbon than α ferrite and is unstable at all temperatures in eutectoid steel. It is soft and non-magnetic.

3. Cementite

It is carbide of iron (Fe_3C) and is formed by definite combination of carbon with iron. It is extremely hard as compared to ferrite and austenite and contains 6.7% carbon and 93.3% iron. High percentage of carbon results in increase in hardness and decrease in ductility.

4. Pearlite

It is the combination of 87.5% ferrite and 12.5% cementite. Steel with 0.8% carbon is wholly pearlite. Hypoeutectic steel is soft due to presence of ferrite and pearlite whereas hypereutectoid steel is hard and brittle due to presence of cementite and pearlite.

5. Martensite

Martensite is a solid solution of iron carbide in α-iron. It is magnetic and has a needle like structure. It is the hardest constituent of Iron carbon system and is the main constituent of hardened steel. It is not as tough as austenite.

6. Sorbite

Sorbite is produced by tempering hardened plain carbon steel. It is finely dispersed pearlite and its properties lies between pearlite and troosite. It is commonly found in heat treated constructional steels.

7. Troostite

It is obtained by quenching tempering martensite plain carbon steel between 200°C and 450°C. It is composed of two equilibrium phases of ferrite and cementite. It is less hard and brittle than martensite.

8. Bainite

It is obtained by isothermal decomposition of austenite. Bainite is obtained in two forms i.e. upper bainite at 400°C and lower bainite at 325°C which is nearly similar to lightly tempered martensite.

3.5. HEAT TREATMENT PROCESSES

1. Annealing
2. Normalizing
3. Hardening
4. Tempering
5. Case hardening

3.5.1. Annealing [UPTU May 2009]

Annealing consists of heating a metal to required temperature, holding it at that temperature for sufficient time to allow internal changes and then cooling at a very slow rate. Objectives of annealing are as follows:

1. To refine the grains and provide structural homogeneity.
2. To soften the metal.
3. To relieve internal stresses.
4. To remove the gases trapped during casting of metal.
5. To produce desired microstructure.
6. To prepare steel for further treatment.
7. To change physical and magnetic properties.

Annealing can be broadly classified as :

(i) Full Annealing
(ii) Process Annealing
(iii) Spheroidise Annealing
(iv) Diffusion Annealing.

(i) Full Annealing

It is high temperature annealing process which removes all structural imperfections by complete recrystallization. The main objectives are to soften the metal, relieve stresses and refine the grain structure.

Full annealing consist of heating the steel to a temperature which is 30°C to 50°C above higher critical point for hypoeutectoid steel and by same temperature range above the lower critical point for hypereutectoid steels. After heating to specified temperature metal is hold at that temperature for sufficient time to allow internal changes in the structure. The metal is now slowly cooled in the furnace. The rate of cooling varies with composition. With slow cooling rate austenite decomposes into ferrite and pearlite in hypoeutectoid steel, and into cementile and pearlite in hypereutectoid steel.

(ii) Process Annealing

It is also known as low temperature annealing process and is used to remove the ill effect of cold working. It softens the metal so that its ductility can be restored for further plastic deformation.

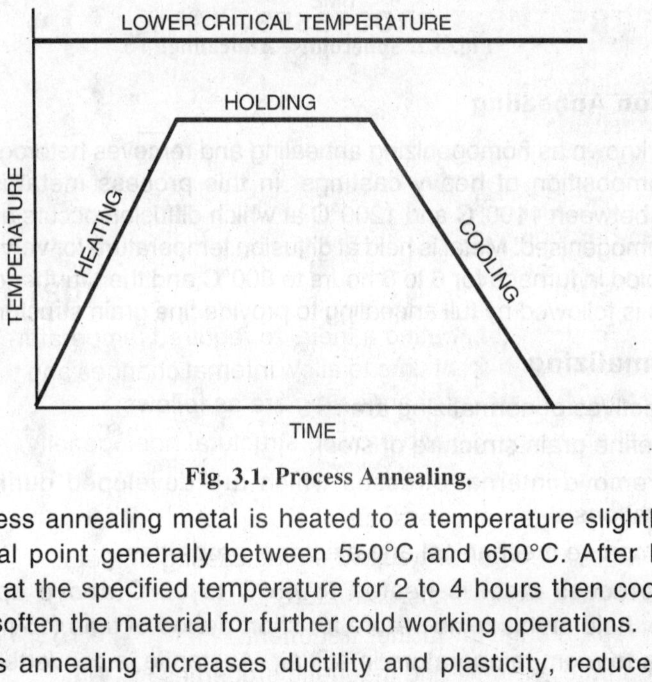

Fig. 3.1. Process Annealing.

In process annealing metal is heated to a temperature slightly below the lower critical point generally between 550°C and 650°C. After heating it is maintained at the specified temperature for 2 to 4 hours then cooled in air or furnace to soften the material for further cold working operations.

Process annealing increases ductility and plasticity, reduces hardness, remove internal stress, and improve machinablity. This process is generally used in sheet and wire industry.

(iii) Spheroidise Annealing

In this process cementite is produced in form of granules and causes all carbides in the steel to agglomerate in form of small globules or spheroids. It consists of heating the metal to a specified temperature slightly above the lower critical point (730°C to 770°C), holding it at this temperature for sufficient period of time and then cooling in furnace to 600°C, followed by cooling in still air to room temperature. This process is used for improving the machinability of high carbon steels, which are difficult to machine.

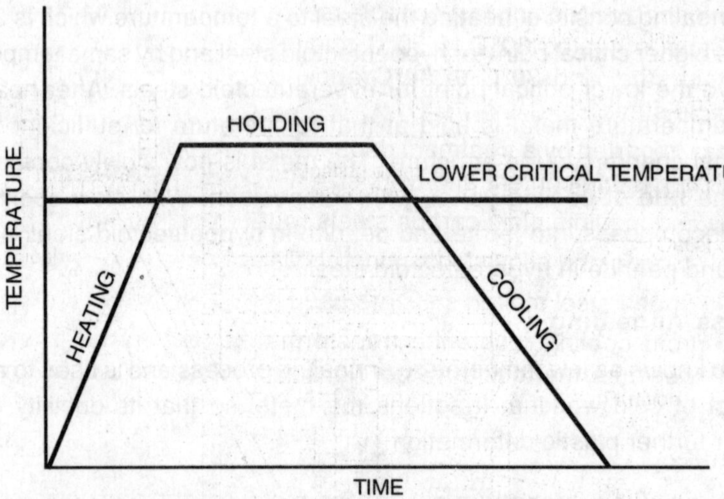

Fig. 3.2. Spheroidise Annealing.

(iv) Diffusion Annealing

It is also known as homogenizing annealing and removes heterogeneity in the chemical composition of heavy castings. In this process metal is heated to temperature between 1100°C and 1200°C at which diffusion occurs and austenite grains are homogenised. Metal is held at diffusion temperature for very short period and then cooled in furnace for 6 to 8 hours to 800°C and then further cooled in air. This process is followed by full annealing to provide fine grain structure.

3.5.2. Normalizing

The objectives of normalizing are :

1. To refine grain structure of steel.
2. To remove internal stresses which are developed during previous operations.
3. To improve mechanical and electrical properties.

In this process steel is heated to 45°C to 50°C above upper critical temperature, held at this temperature for short period of time (15 minutes) and cooled down to room temperature in still air. Austenite is converted into ferrite and pearlite for hypoeutectoid steel, into pearlite for eutectoid steel, and into pearlite and cementite for hypereutectoid steel.

Features

1. The steel obtained is harder and stronger but less ductile than annealed steel.
2. It also improves yield point, impact strength, and ultimate tensile strength of steel.
3. It is applied to castings and forgings.

3.5.3. Hardening

Hardening process consists of heating the steel to a temperature above its critical point, holding it at that temperature for sufficient time depending upon its thickness and then quenching in a suitable medium such as water, oil or molten salt bath. Steel is heated 30°C to 50°C above upper critical temperature for hypoeutectoid steel and 30°C to 50°C above lower critical temperature for hypereutectoid steel.

Hardness imparted by a treatment depends upon number of factors such as rate of cooling, carbon content and work size. For rapid cooling of

(i) Low and medium plain carbon steels water or brine solution is used.

(ii) High carbon and alloy steels mineral oils are used.

(iii) High speed steel molten salts are used.

Due to rapid cooling, austenite transforms into a very hard and brittle constituent known as martensite. Steel hardness increases with formation of martensite.

Hardenability

Hardenability indicates the ability of the metal to attain uniform hardness throughout its mass or varying hardness to some depth. It determines depth and distribution of hardness in metal due to quenching. Fully hardened articles will have same properties throughout its cross-section.

Some factors which affect hardenability are

1. Method of quenching and quenching medium.
2. Composition of steel.
3. Shape and size of job.
4. Grain size.
5. Quenching rate.
6. Surface condition.
7. Heating rate and time.

3.5.4. Tempering

It is defined as a process in which hardened steel is heated to a temperature below the lower critical temperature to transform the hard and brittle martensite into ferrite and cementite. In this process ductility and toughness is achieved at the cost of hardness and strength. Tempering can be classified as

(i) Low temperature tempering (150°C to 250°C)

(ii) Medium temperature tempering (350°C to 450°C)

(iii) High temperature tempering (500°C to 650°C)

Low tempering temperature in the range of 150°C to 250°C relieves the internal stresses. As the temperature is increased from 200°C to 450°C, ductility and toughness is achieved at the cost of hardness. Further increase of temperature to 550°C results in complete elimination of internal stresses and more increase in ductility. Tempering above 550°C is not useful because hardness is reduced without any remarkable increase in ductility.

3.5.5. Case Hardening

Case hardening produces a hardened case or surface over a tough, shock resistant core. It is required in gears, ball bearings, camshafts, etc. to provide wear resistant faces to interior part which is soft and ductile to absorb any shock. Case hardening is done in two stages :

(i) Carburising

(ii) Hardening.

In carburising, carbon is diffused into low carbon steel surface by process of diffusion from carbon monoxide gas at temperature of 870°C to 950°C. When the hot metal is quenched the outer surface hardens whereas the core being low carbon steel remains soft and ductile.

Carburising can be done by using three methods :

(i) Pack carburising.

(ii) Liquid carburising.

(iii) Gas carburising.

(1) Pack Carburising

Iron parts which are to be carburised are placed in a heat resistant box made of cast iron or steel. This box contains mixture of carbonaceous materials such as charcoal, wood, tar, charred leather, etc. Box is properly sealed and gradually heated in furnace between 900°C to 950°C, slightly above the critical range. The box is held at that temperature for 3 to 15 hours so that carbon is absorbed in the metal to the required depth and outer surface is converted into high carbon steel. After this metal is cooled slowly in the box and removed. The depth upto which carbon is enriched depeds upon holding time.

The metal piece is again reheated slightly above its critical point (above 915°C) followed by quenching in oil, water or brine to refine the structure and make the core soft and tough. The metal piece is again reheated to about 750°C and quenched to increse the hardens of case.

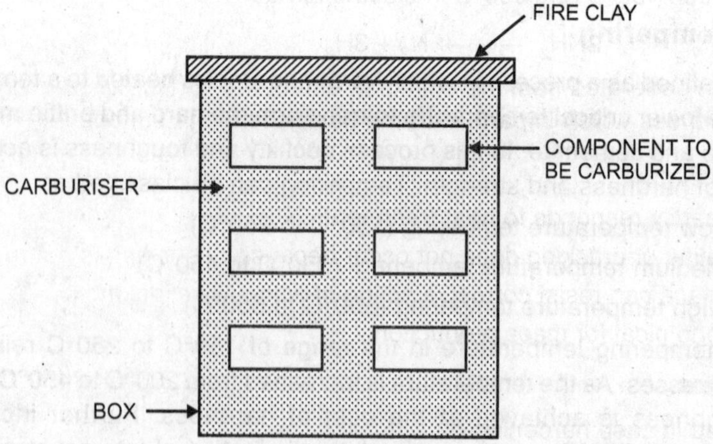

Fig. 3.3. Pack carburising.

(2) Liquid Carburising

Process of liquid carburising is similar to pack carburising but in this process liquid medium is used inplace of carbonaceous material. The parts which are to carburised are placed in liquid salt baths of molten salt containing sodium carbonate, sodium chloride and silicon carbide for sufficient time. Temperature in liquid carburising varies between 840°C to 920°C. Component is then removed from bath and quenched in oil, water or brine. Some advantages of liquid carburising are :

(i) Large depth of penetration is possible.
(ii) Very low deformation of component.
(iii) Uniform heating of component.
(iv) Easy to carburise a wide range of parts.

(3) Gas Carburising

In gas carburising carbon monoxide reacts with surface of steel and results in direct and rapid absorption of carbon in surface. In this process, component is suspended in gas furnace from hooks and heated at about 900°C for 3 to 4 hours in atmosphere of gases rich in hydrocarbons such as methane, butane, propane, etc. to absorb carbon. After soaking component is directly quenched without exposing to atmosphere to make the surface of component hard with excellent surface finish. Advantages of gas carburising are :

(i) Very high surface finish is obtained.
(ii) Fast process as compared to pack carburising.
(iii) Less space is required.
(iv) Accuracy of desired case depth.

3.5.6. Nitriding

It is defined as a process in which steel surface is saturated with nitrogen by holding it in atmosphere of ammonia for prolonged periods in the temperature range between 480°C and 650°C in electric furnace.

$$2NH_3 \longrightarrow N_2 + 3H_2$$

Nitrogen generated from the ammonia enters the surface of steel and form very hard nitrides which imparts extreme hardness to the surface.

Advantages

(i) It readily responds to heat treatment.
(ii) Scaling or cracking does not occur because quenching is not required.
(iii) Surface can resist corrosive action of water, alkalies, etc.
(iv) Economical for mass production.

Applications

It is used in case hardening of

(i) Automobile engine parts, gears, clutches, etc.
(ii) Ball and roller bearings.
(iii) Mandrel, drawing dies, etc.

3.5.7. Cyaniding

It is defined as a process in which nitrogen and carbon are added to surface of steel by immersing it in a molten cyanide bath maintained at 850°C to 950°C.

After soaking for required time the steel is quenched in water or oil to obtain a hard outer surface. During heating nitrides are formed and during cooling carbides are formed on the surface. Cyaniding process provide hardness generally upto depth of 0.5 mm.

It imparts bright surface finish and increased fatigue limit. It is generally used for low and medium carbon steel parts such as gears, brake, cam, sleeves, shafts, etc.

3.5.8. Induction Hardening

In this process surface to be hardened is placed between inductor block and a high frequency low voltage current (2000 cycles per sec) is passed through the block. Inductor block induces current in the surface of the metal and heats it by induced eddy current. Heated surface is quenched simultaneously with water by numbers of holes provided in inductor block. To increase the depth of penetration the hardening temperature is maintained above 768°C. It is used for hardening the surfaces of camshafts, crankshafts, gears, axles, brake drums, etc.

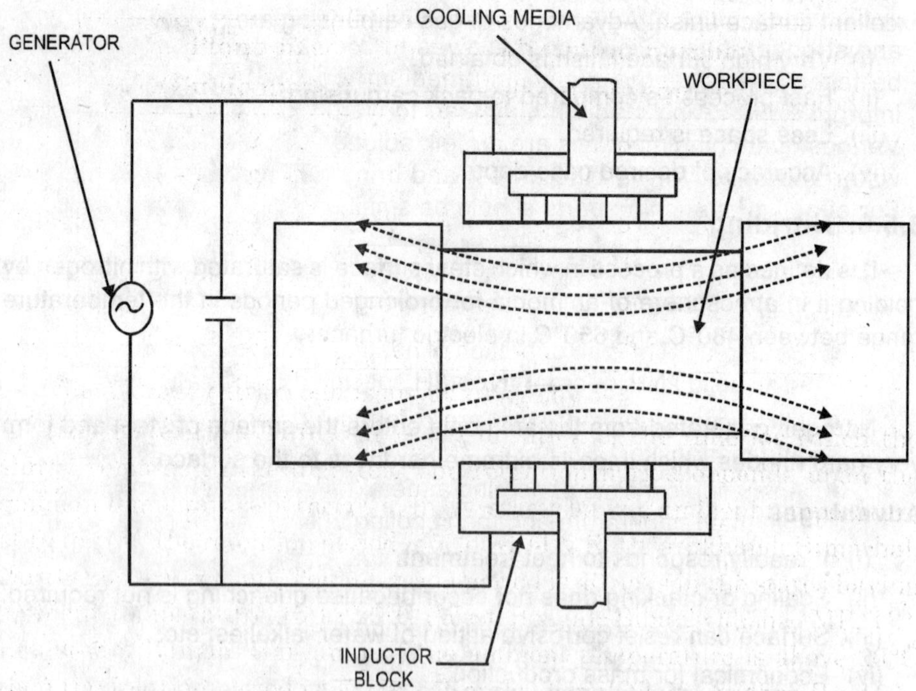

Fig. 3.4. Block Diagram of Induction Hardening.

3.5.9. Flame Hardening

This hardening is used to harden the particular portion of job. The portion to be hardened is heated with oxyacetylene flame above its critical temperature followed by quenching by means of spray of water. In flame hardening stresses are not developed due to localized heating and chances of cracking and distortion are reduced. It is used for hardening gears, wheels, mill rolls, etc. Main advantage is low cost of equipment, less time for heating and less distortion as compared to other process.

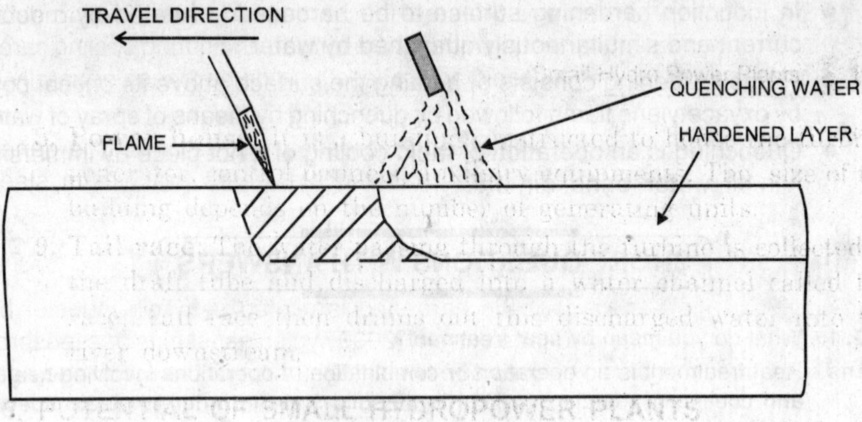

Fig. 3.5. Flame Hardening.

3.6. QUENCHING

Quenching is an operation of rapid cooling of a hot piece by immersing it in oil, water, brine, air etc. This rapid cooling makes the steel much harder and stronger than it would be if allowed to cool slowly. Enough care should be taken in selecting a quenching medium for a particular metal because improper quenching medium may lead to warping or cracking of the piece. Various quenching mediums are air, oil, solution of water and sulphuric acid, water and brine. Air has the lowest and brine has the highest cooling rate. For effective quenching parts should be small and thin in cross-section.

SUMMARY

- Heat treatment is an operation of heating and cooling of metals or alloys in its solid state in order to obtain specific properties.
- Micro constituents of steel are α-ferrite, austenite, cementite, pearlite and martensite.
- Annealing consists of heating a metal to required temperature, holding it at that temperature and then cooling at a very slow rate.
- Annealing is broadly classified as full annealing, process annealing, spheroidise annealing and diffusion annealing.
- In normalizing steel is heated to temperature 45°C to 50°C above upper critical temperature, held at this temperature for short period and cooled down to room temperature in still air.
- Hardening consists of heating the steel to a temperature above its critical point, holding it at that temperature and then quenching in suitable medium.
- Hardenability indicates the ability of the metal to attain uniform hardness throughout its mass or varying hardness to some depth.
- Tempering is done to transform martesite into ferrite and cementite.
- Case hardening produces a case or hardened surface over a tough shock resistance core.
- Nitriding is a process of saturating steel surface with nitrogen.

- In induction hardening surface to be hardened is heated by induced current and simultaneously quenched by water.
- Flame hardening consists of heating the surface above its critical point by oxyacetylene flame followed by quenching by means of spray of water.
- Quenching is an operation of rapid cooling of a hot piece by immersing it in oil, water, brine, air, etc.

SHORT QUESTIONS WITH ANSWERS

Q. 1. What do you mean by heat treatment?

Ans. Heat treatment is an operation or combination of operations involving heating and cooling of metals or alloys in its solid state in order to obtain specific properties.

Q. 2. List any four microscopic constituents of iron and steel.

Ans. (1) α Ferrite

(2) Austenite

(3) Cementite

(4) Pearlite

Q. 3. Give the temperature at which α-iron, β-iron, γ-iron, and δ-iron exists.

Ans. α-iron : below 768°C

β-iron : between 768°C and 900°C

γ-iron : between 900°C and 1405°C

δ-iron : between 1405°C and 1535°C

Q. 4. Why normalizing is done?

Ans. Normalizing is done

(i) To refine grain structure of steel.

(ii) To remove internal stresses.

(iii) To improve mechanical and electrical properties.

Q. 5. What is the effect of rapid cooling on micro constituents of steel in hardening process?

Ans. Due to rapid cooling, austenite transforms into a hard and brittle constituent known as martensite.

Q.6. List four factors which affect hardenability.

Ans. (a) Method of quenching and quenching medium.

(b) Composition of steel

(c) Shape and size of job

(d) Grain size.

Q. 7. Why the tempering above 550°C is not useful?

Ans. Tempering above 550°C reduces hardness without any remarkable increase in ductility.

Q. 8. Why case hardening is required?

Ans. To impart wear resistance with tough core is not possible in homogeneous composition of steel so surface layer of specimen are hardened by keeping core soft and ductile.

Q. 9. What is flame hardening?

Ans. Flame hardening consists of heating the surface above its critical point by oxyacetylene flame followed by quenching by means of spray of water.

Q. 10. Define quenching.

Ans. Quenching is an operation of rapid cooling of a hot piece by immersing it in oil, water, brine, air, etc.

Q. 11. What is the basic aim of tempering. **[UPTU Dec. 2011]**

Ans. The basic aim of tempering is to relieve resuidal stress, improve ductility and toughness, reduce hardness.

REVIEW QUESTIONS

1. What do you mean by heat treatment? Explain the objectives of heat treatment.
2. What is annealing process and explain various types of annealing operations.
3. What is normalizing? Why is it done?
4. Define hardening and why is it required?
5. Explain hardenability. What is the difference between hardness and hardenability?
6. What is tempering and explain the different stages of tempering.
7. Why is case hardening done? Explain in detail.
8. Write short notes :
   (a) Nitriding
   (b) Cyaniding
9. What do you mean by case hardening? Explain different method of case hardening. **[UPTU Dec. 2010]**
10. Explain in detail
    (a) Induction hardening
    (b) Flame hardening
11. Write short notes on annealing and normalizing of carbon steels. **[UPTU Dec. 2009]**

SELF TEST QUESTIONS

A. State Whether True or False

1. Austenite has much greater solid solubility for carbon than α-ferrite.
2. Martensite is the hardest constituent of iron carbon system.
3. Steel hardened by normalizing is harder and stronger but less ductile than annealed steel.
4. Low percentage of carbon results in increase in hardness and decrease in ductility.
5. Annealing is done to impart hardness.

B. Fill in The Blanks

1. Austenite is a solid solution of .... and .... in gamma iron.
2. Martensite is a solid solution of ....in α iron.
3. Full annealing consists of heating the steel 30°C to 50°C above............ for hypoeutectoid steel.
4. α Ferrite is a solid solution of carbon in ....
5. Tempering above .... is not useful.

C. Multiple Choice Questions

1. Pearlite is a combination of
(a) 87.5% ferrite and 12.5% cementite.
(b) 12.5% ferrite and 87.5% cementite.
(c) 50% ferrite and 50% cementite.
(d) 25% ferrite and 75% cementite.

2. Steel with 0.8% carbon is wholly
(a) Pearlite (b) Austenite
(c) Martensite (d) Cementite

3. Heat treatment is done to
(a) Relieve internal stresses (b) Improve mechanical properties
(c) To remove trapped gases
(d) All of the above

4. For full annealing of hypereutectoid steel the steel is heated 30°C to 50°C above
(a) Higher critical point
(b) Lower critical point
(c) Both
(d) None

5. Factors which affect hardenability are
(a) Method of quenching (b) Composition of steel
(c) Grain size (d) All of the above.

6. Process of adding nitrogen and carbon to surface is
(a) Cyaniding (b) Nitriding
(c) Quenching (d) All of the above

7. Process in which surface layers of specimen are hardened by keeping core soft and ductile is known as
(a) Tempering (b) Normalizing
(c) Annealing (d) Case hardening

8. Rapid cooling of steel makes it
(a) Stronger and harder (b) Soft and ductile
(c) Soft and tough (d) None

9. Cementite contains
(a) 93.3% carbon and 6.7% iron
(b) 6.7% carbon and 93.3% iron
(c) 2% carbon and 98% iron
(d) 50% carbon and 50% iron

10. Process consisting of heating the surface above its critical point by oxy-acetylene flame followed by quenching with spray of water is
(a) Induction hardening (b) Flame hardening
(c) Gas hardening (d) Quenching

11. Purpose of heat treatment is
(a) To change and refine grain size
(b) Relieve internal stress
(c) Modify electrical and magnetic property
(d) All of these

12. Which constituent has bcc structure
(a) α ferrite (b) Austenite
(c) Cementite (d) Bauxite

13. α ferrite is solid solution of carbon in
(a) Gamma iron (b) Beta iron
(c) α iron (d) None of these

14. Which constituent has 87.5% ferrite and 12.5% cementite
(a) Austenite (b) Martensite
(c) Sorbite (d) Pearlite

15. Hardest constituent of steel is
(a) Austenite (b) Cementite
(c) Martensite (d) α ferrite

16. Constituent obtained by isothermal decomposition of austenite is
(a) α ferrite (b) Bainite
(c) Cementite (d) Pearlite

17. Which annealing is known as low temperature annealing
(a) Process annealing (b) Full annealing
(c) Spheroidise annealing (d) Diffusion annealing

18. What is the temperature range for process annealing of metal
(a) 200 to 310°C (b) 350 to 450°C
(c) 550 to 650°C (d) 450 to 550°C

19. In spheriodise annealing metal is heated
(a) Above higher critical temperature
(b) Below lower critical temperature
(c) Above lower critical temperature
(d) Below higher critical temperature

20. Annealing used to remove hetrogeneity in chemical composition of heavy casting is
(a) Diffusion annealing (b) Full annealing
(c) Process annealing (d) Spheroidise annealing

21. Temeprature range for full annealing is
(a) 10° to 30°C (b) 30° to 50°C
(c) 50° to 70°C (d) 70° to 90°C

22. In which process steel is heated above upper critical temperature
(a) Tempering (b) Spheroidise annealing
(c) Normalizing (d) Process annealing

23. During the hardening of high speed steel which coolant is used for rapid cooling
(a) Water (b) Mineral oil
(c) Lime solution (d) Molten salts

24. Due to rapid cooling, austenite transforms into a very hard and brittle material known as
(a) Martensite (b) Pearlite
(c) Ferrite (d) None of these

25. Temperature range for high temperature tempering is
(a) 150° to 250°C (b) 350° to 450°C
(c) 500° to 650°C (d) 650° to 750°C

26. In tempering process steel is heated
    (a) Above the upper critical temperature
    (b) Above the lower critical temperature
    (c) Below the upper critical temperatrure
    (d) Below the lower critical temperature
27. Process used to produce a hardened case over a tough, shock, resistant core is called as
    (a) Case hardening (b) Normalizing
    (c) Tempering (d) Quenching
28. In which process carbonaceous material is used
    (a) Liquid carburising (b) Gas carburising
    (c) Pack carburising (d) Nitriding
29. Which gas is used to harden the surface in nitriding process
    (a) Oxygen (b) Carbon dioxide
    (c) Carbon monoxide (d) Ammonia
30. Process in which nitrogen and carbon are added to surface of steel by immersing it in a molten cyanide bath
    (a) Nitriding (b) Cyaniding
    (c) Normalizing (d) Quenching
31. In which process oxyacetylene flame is used to harden the particular portion of job
    (a) Induction hardening (b) Quenching
    (c) Flame hardening (d) Nitriding
32. Which coolant has highest cooling rate **[UPTU Dec. 2010]**
    (a) Brine (b) Air
    (c) Water (d) Sulphuric acid

D. Match the Following

1.

(A)	(B)
(i) Ferrite	(a) FCC structure
(ii) Austenite	(b) BCC structure
(iii) Cementite	(c) Needle like structure
(iv) Martensite	(d) Carbide of iron

2.

(A)	(B)
(i) Full annealing	(a) Below the lower critical point
(ii) Process annealing	(b) Above the lower critical point
(iii) Spheroidise annealing	(c) Homogenizing annealing
(iv) Diffusion annealing	(d) High temperature annealing process

3.

(A)	(B)
(i) Nitriding	(a) High frequency low voltage current
(ii) Cyaniding	(b) Atmosphere of ammonia
(iii) Flame hardening	(c) Nitrogen and carbon
(iv) Induction hardening	(d) Oxyacetylene flame

4.

(A)	(B)
(i) α-iron exists	(a) Between 900°C and 1405°C
(ii) β-iron exists	(b) Between 1405°C and 1535°C

(iii) γ-iron exists — (c) Between 768°C and 900°C
(iv) δ-iron exists — (d) Below 768°C

5.

(A)	(B)
(i) Pearlite	(a) Solid solution of carbon in α-iron
(ii) Cementite	(b) Solid solution of iron carbide in α-iron
(iii) α-ferrite	(c) Combination of ferrite and cementile
(iv) Martensite	(d) 6.7% carbon and 93.3% iron

6.

(A) Name of treatment	(B) Media used
(i) Pack carburising	(a) Ammonia gas
(ii) Gas carburising	(b) Molten cyanide bath
(iii) Cyaniding	(c) Carbonaceous component
(iv) Nitriding	(d) Methane

7.

(A)	(B)
(i) Annealing	(a) Cool the workpiece
(ii) Nitriding	(b) Increase surface hardness
(iii) Quenching	(c) Improve ductility

ANSWERS

A. True or False

1. True **2.** True **3.** True **4.** False **5.** False

B. Fill in the Blanks

1. Ferrite and iron carbide **2.** Iron carbide **3.** Higher critical point
4. α iron **5.** 550°C

C. Multiple Choice Questions

1. (a) **2.** (a) **3.** (d) **4.** (b) **5.** (d)
6. (a) **7.** (d) **8.** (a) **9.** (b) **10.** (b)
11. (d) **12.** (a) **13.** (c) **14.** (d) **15.** (c)
16. (b) **17.** (a) **18.** (c) **19.** (c) **20.** (a)
21. (b) **22.** (c) **23.** (d) **24.** (a) **25.** (c)
26. (d) **27.** (a) **28.** (c) **29.** (d) **30.** (b)
31. (c) **32.** (a)

D. Match the Following

1. (i)—(b), (ii)—(a), (iii)—(d), (iv)—(c) **2.** (i)—(d), (ii)—(a), (iii)—(b), (iv)—(c)
3. (i)—(b), (ii)—(c), (iii)—(d), (iv)—(a) **4.** (i)—(d), (ii)—(c), (iii)—(a), (iv)—(b)
5. (i)—(c), (ii)—(d), (iii)—(a), (iv)—(b) **6.** (i)—(c), (ii)—(d), (iii)—(b), (iv)—(a)
7. (i)—(c), (ii)—(b), (iii)—(a)

❑❑❑

CHAPTER 4

Non-Ferrous Materials

INSIDE THIS CHAPTER

4.1. INTRODUCTION

Nonferrous materials are those, which contain material other than iron as their main constituent. Most widely used non-ferrous materials are aluminium, copper, zinc, magnesium, nickel, etc. Generally non-ferrous materials have high cost as compared to ferrous materials but they are extensively used due to their excellent and wide range of properties, ease of fabrication, good electrical and thermal conductivity, high corrosion resistance, low density, softness, etc.

4.2. ALUMINIUM

Aluminum is very abundant metal in earth and is available in various forms of oxides, sulphates, phosphates etc. Bauxite is the commercially available ore of aluminum and is a mixture of monohydrate ($Al_2O_3.H_2O$) and tri-hydrate ($Al_2O_3.3H_2O$). Aluminium is extracted by electrolysis of the mixture of purified bauxite dissolved in fused cryolite.

Properties of Aluminium

1. Light in weight, malleable and ductile.
2. Excellent thermal and electrical conductivity.
3. Low specific gravity i.e. 2.7.
4. Non-magnetic and non-toxic.
5. Melting point is 658°C and boiling point is 2056°C.
6. Low tensile strength.
7. High resistance to corrosion.

Uses

1. In manufacture of domestic utensils and heat conducting appliances.
2. In electrical conductor and busbars.
3. In casting of steel.
4. In manufacture of containers, window frames, structural members, furniture, etc.
5. Aluminium foils are used for packing of chocolates, food etc.
6. Used in thermit welding.
7. In collapsible tubes for paste.
8. In aircraft and automobile components.

4.3. ALUMINIUM ALLOYS

Main aluminium alloys are as follows :

1. Duralumin
2. Y- alloy
3. Magnalium
4. Hindalium

4.3.1. Duralumin

It contains 3.5% to 4.5% copper, 0.4% to 0.7% manganese, 0.4% to 0.7% magnesium, 0.5% iron or silica and rest is aluminium.

Properties of Duralumin

1. Excellent casting and forging properties.
2. High tensile strength.
3. High electrical conductivity.
4. High machinability.
5. Melting point 650°C, specific gravity 2.8.

Uses

It is widely used in manufacture of aircraft parts, automobile parts, sheet, tubes, rivets, forging, cables, etc.

4.3.2. Y-Alloys

It contains 4% copper, 1.2% to 1.7% manganese, 1.8% to 2.3% nickel and rest is aluminium. It is used for piston cylinder head, rods, etc.

Properties of Y-Alloys

1. Better strength than duralumin at high temperature.
2. Can be easily cast and hot worked.

4.3.3. Magnalium

It contains 1 to 5.5% magnesium, 1.75% copper and rest is aluminium. It is light in weight and has good machinability. It is mainly used in aircraft and automobile parts.

4.3.4. Hindalium

It is an alloy of aluminium, magnesium, chromium, manganese, etc. and is the trade name of the aluminium alloy produced by Hindustan Aluminium Corporation Limited, Renukot, U.P.

4.4. COPPER [UPTU DEC. 2008]

Copper occurs in combined state under the earth and is one of the most widely used non ferrous materials. Its main ore are cuprite (Cu_2O), copper glance(Cu_2S) and copper pyrites ($CuFeS_2$). To extract copper, ores are cleaned, crushed and calcined in a reverberatory furnace. Calcined ores are mixed with silica and coke, and smelted in a blast furnace. This molten metal is oxidized in Bessemer convertor to give blister copper. Pure copper is then obtained by melting blister copper in reverberatory furnace followed by electrolysis. It is reddish brown in colour and is a good conductor of electricity. It is used in making coins, electric cables, utensils and manufacturing of brass and bronze alloys.

Properties of Copper

1. Its melting point is 1083°C and boiling point is 2300°C.
2. It gives green coating when brought in contact with moist air.
3. It has good corrosion resistant.
4. It has high electrical and thermal conductivity.

Copper alloys can be broadly classified as :

(i) Brass – An alloy of copper and zinc.

(ii) Bronze – An alloy of copper and tin.

4.4.1. Brass

Brass is an alloy of copper and zinc. It generally contains 60 to 90% copper and 10 to 40% zinc. Brass is the most widely used copper alloy. It can be rolled into sheets, drawn into wires, cast into moulds and can be easily fabricated by processes like spinning and Y-chromium. According to the percentage of copper and zinc, brass is of following types:

1. **Low Brass :** It contains 80 % copper and 20 % zinc and is used in forming and musical equipments.
2. **Red Brass :** It contains 85 % copper and 15% zinc and is used in plumbing pipe, electrical sockets, radiator core, etc.
3. **White Brass :** It contains 10% copper and 90% zinc and generally used in ornaments.
4. **Yellow Brass :** It contains 60% copper and 40% zinc and is used in rivets, tubes, valve stems, etc.
5. **Cartridge Brass :** It contains 70% copper and 30% zinc and is used in cartridges, tubes, springs, etc.

Properties of Brass

1. They have high strength and hardness.
2. Less thermal and electrical conductivity as compared to copper.
3. High corrosion resistance.

4.4.2. Bronze [UPTU Dec. 2008]

Bronze is an alloy of copper and tin. It generally contains 75 to 95% copper and 5 to 25% tin. It has better mechanical and corrosion properties as compared to brass. It can be rolled into sheets, wire and rods. It is used in making utensils, pump linings, bushes, bearings and hydraulic fittings. Main types of Bronze are as follows :

1. **Phosphor Bronze :** It contains 9% to 10% tin, 0.1% to 0.3% phosphorus and remaining is copper. It has high strength and toughness, good bearing capacity, high elasticity, and resistance to corrosion. It is used in bearings, gears, pumps parts, springs, etc.
2. **Aluminium Bronze :** It does not contain tin and is an alloy of 90% copper and 10% aluminium. It has good strength and high corrosion resistance. It is used in condensers, heat exchangers, gears, cams and rollers, etc.
3. **Beryllium Bronze :** It contains 98% copper and 2% beryllium. It has high yield point and fatigue limit. It is used in springs, cams and bearings.
4. **Silicon Bronze :** It contains 96% copper, 3% silicon, 0.5% manganese, 0.5% iron. It has high strength and toughness. These are used in boilers, tubes, pumps, etc.
5. **Manganese Bronze :** It contains 55% copper, 38% zinc, 5% manganese and 2% tin. It is used in bushes, rods, ship propellers, etc.
6. **Gun Metal :** It contains 88% copper, 10% tin and 2% zinc. It is used in nut, bolts, bearings, casting, guns, etc.
7. **Bell Metal :** It contains 80% copper and 20% tin. It is used in casting of bells.

Properties of Bronze

1. It is comparatively hard and resist surface wear.
2. It has antifriction bearing properties.
3. It has better corrosion resistance than brass.

4.5. LEAD

Lead is soft and heavy material, which is found in combined form with sulphide known as galena. Other ores of lead are Anglesite, cerussite, mimetesite, etc. Lime, silica, coke and metallic iron are added to the ores and fed into a blast furnace, ore get melted and impure lead is obtained which is further fed to reverberatory furnace to reduce impurities.

Properties of Lead

1. It is poisonous material.
2. It has high density and heavy weight.
3. Low melting point (327°C)
4. It can be easily welded and soldered.
5. It is soft and malleable.

Uses of Lead

1. It is used in bearing materials.
2. Storage batteries.
3. Making alloys like brass, bronze etc.
4. Lining to prevent corrosion.

4.6. NICKEL

It is a lustrous white material, extracted from sulphide ores and has good corrosion resistance.

To obtain nickel first of all the ore is roasted. Limestone and quartz are added to it and mixture is then melted in blast furnace. Molten mixture of nickel and copper is collected from bottom of furnace and led to Bessemer converter. After treatment in converter it is treated with sulphuric acid to obtain metallic nickel.

Properties of Nickel

1. It is malleable and ductile.
2. Its melting point is 1455°C and boiling point is 2730°C.
3. It is a ferromagnetic material.

Uses of Nickel

1. For coating of other metals like copper, brass, etc.
2. As alloying element.
3. Manufacture of crucible, domestic utensils, etc.

4.7. ZINC

It is found in combined form in nature as zinc blende (ZnS), zincite (ZnO) and calamine ($ZnCO_3$).

To obtain zinc, some carbonaceous material is added to zinc ore and fed into electric furnace. Impurities are removed as gases and zinc is liberated in vapour form. These vapours are then condensed to get the metallic zinc.

Properties of Zinc

1. It is bluish white in colour.
2. Its melting point is 419.5°C and boiling point is 920°C.
3. It has high resistance to atmospheric corrosion.
4. It has electropositive character.

Uses of Zinc

1. In production of alloys.
2. In electric cells for galvanizing iron.
3. In protective covering and coating.
4. In stampings and die casting.

4.8. TIN

It is a nontoxic, shining, white metal. It is generally obtained from tinstone (SnO_2) ore. To extract tin, the ore is crushed and washed. It is then calcined and smelted in furnace with anthracite coal and sand to obtain crude tin. This crude tin is remelted and refined further in reverberatory furnace to get commercially pure tin. It is finally refined by electrolytic deposition method to obtain highly pure tin.

Properties of Tin

1. Its melting point is 232°C and boiling point is 2260°C.
2. It has high ductility and can easily be cold worked.
3. It has good corrosion resistance.
4. Good resistance to acid attack.

Uses of Tin

1. For coating on other metals.
2. As an alloying element in solder, bronze, bell metal etc.
3. In making the foil for moisture proof packing.

SUMMARY

- Nonferrous materials are those, which contain material other than iron as their main constituent.
- Aluminium is extracted by electrolysis of the mixture of purified bauxite dissolved in fused cryolite.
- Main alloys of aluminium are Duralumin, Y-alloy, Magnalium and Hindalium.
- Brass generally contains 60 to 90% copper and 10 to 40% zinc.
- Bronze generally contains 75 to 95% copper and 5 to 25% tin.
- Lead is soft and heavy metal and its chief ore is galena.
- Nickel is a lustrous white material which is extracted from sulphide ores.
- Zinc is a bluish white material obtained from zinc blende (ZnS), zincite (ZnO) and calamine ($ZnCO_3$).
- Tin is a nontoxic, shining, white material, generally obtained from tinstone (SnO_2) ore.

SHORT QUESTIONS WITH ANSWERS

Q. 1. How aluminium is extracted from bauxite?

Ans. Aluminium is extracted by electrolysis of the mixture of purified bauxite dissolved in fused cryolite.

Q. 2. Why aluminium is used in aircraft and automobile components?

Ans. Aluminium is light in weight, malleable and ductile. It has low specific gravity i.e. 2.7.

Q. 3. Name any four aluminium alloys. **[UPTU May 2012]**

Ans. (1) Duralumin
(2) Y-alloys
(3) Magnalium
(4) Hindalium.

Q. 4. Name any two ores of copper.

Ans. Cuprite (Cu_2O) and Copper glance (Cu_2S).

Q. 5. What do you mean by Brass and Bronze?

Ans. Brass is an alloy of copper and zinc whereas bronze is an alloy of copper and tin.

Q. 6. Give the composition of Gun Metal.

Ans. Copper-88%, Zinc-2%, and Tin-10%.

Q. 7. What are the uses of lead?

Ans. It is used in bearing materials and storage batteries.

Q. 8. Name any two ore of Zinc.

Ans. Zinc blende (ZnS) and Calamine ($ZnCO_3$).

Q. 9. List the alloying elements of duralumin. **[UPTU Dec. 2009]**

Ans. Copper (Cu), Magnesium, Maganese, Iron, Aluminium.

REVIEW QUESTIONS

1. What are non-ferrous materials?
2. Why non-ferrous materials are extensively used inspite of high cost?
3. Give the properties and uses of aluminium.
4. Give the properties and uses of aluminium alloys.
5. Explain brass and bronze with their properties and applications.
6. What are the major uses of copper ? What are the alloying elements in brass and bronze? **[UPTU Dec. 2008]**
7. Name three alloys each of ferrous and non-ferrous metals. Write their applications. **[UPTU May 2010]**
8. List various type of brasses. Also give their applications. **[UPTU May 2011]**

SELF TEST QUESTIONS

A. State Whether True or False

1. Galena is an alloy of lead.
2. Lead is a toxic material.
3. Bronze is an alloy of copper and zinc.
4. Tin is obtained from tinstone ore.
5. Aluminium is nonmagnetic and nontoxic.

B. Multiple Choice Questions

1. Which of the following is an ore of aluminium?

   (a) Tinstone (b) Calamine

   (c) Bauxite (d) Galena

2. Galena is an ore of

   (a) Copper (b) Lead

   (c) Aluminium (d) Tin

3. Which of the following shows electropositive character?
   (a) Copper (b) Lead
   (c) Zinc (d) Tin
4. Alloys of copper and zinc are known as
   (a) Bronze (b) Brass
   (c) Calamine (d) Tinstone
5. Bronze is an alloy of
   (a) Aluminium and zinc (b) Aluminium and nickel
   (c) Copper and tin (d) Copper and zinc.
   [UPTU Dec. 2008; May 2010]
6. Gun metal contains
   (a) 88% copper, 10% tin, 2% zinc.
   (b) 96% copper, 3% silicon, 0.5% manganese.
   (c) 90% copper and 10% aluminium
   (d) 90% copper and 10% tin.
7. Which of the following metal is used in thermit welding?
   (a) Copper (b) Lead
   (c) Aluminium (d) Tin
8. Which of the following is not a type of brass?
   (a) Cartridge Brass (b) Low brass
   (c) Gun metal (d) Red brass
9. Bell metal contains
   (a) 20% copper and 80% tin.
   (b) 80% copper and 20% tin.
   (c) 98% copper and 2% beryllium.
   (d) 90% copper and 10% aluminium.
10. Which metal is used in storage batteries?
    (a) Lead (b) Tin
    (c) Zinc (d) Steel
11. Bauxite is an ore of
    (a) Aluminium (b) Copper
    (c) Zinc (d) Magnesium
12. Alloy of aluminium is
    (a) Gun metal (b) Y alloy
    (c) Bell metal (d) Phosphor bronze
13. Which metal is used for making foils for packing
    (a) Copper (b) Brass
    (c) Aluminium (d) Bronze
14. Copper ore is
    (a) Cryolite (b) Cuprite
    (c) Galena (d) Bauxite

15. Bronze is a _____ alloy

(a) Aluminium (b) Lead

(c) Copper (d) Zinc

16. Brass is an alloy of

(a) Copper and zinc (b) Copper and tin

(c) Zinc and tin (d) Aluminium and tin

17. Brass contain ______ copper

(a) 40 to 60% (b) 60 to 90%

(c) 20 to 40% (d) 25 to 55%

18. Bell metal contains

(a) Cu and Zn (b) Copper and tin

(c) Tin and zinc (d) Aluminium and copper

19. ______ in an alloy of copper and tin

(a) Brass (b) Lead

(c) Bronze (d) Duralumin

20. Lead ore is ______

(a) Anglesite (b) Bauxite

(c) Cryolite (d) Cuprite

21. Lead is found in combined form with sulphide known as ______

(a) Cuprite (b) Galena

(c) Bauxite (d) None of these

22. Zinc ore is

(a) Calamine (b) Zinc blende

(c) Zincite (d) All of these

C. Match the Following

1.

(A)	(B)
(i) Aluminium	(a) Galena
(ii) Copper	(b) Bauxite
(iii) Lead	(c) Calamine
(iv) Zinc	(d) Cuprite

2.

(A)	(B)
(i) Brass	(a) 88% copper, 10% tin and 2% zinc
(ii) Aluminium bronze	(b) Alloy of copper and zinc
(iii) Gun metal	(c) 80% copper and 20% tin
(iv) Bell metal	(d) 90% copper and 10% aluminium

3.

(A)	(B)
(i) Y-alloy	(a) Melting point is 1455°C
(ii) Gun metal	(b) Poisonous material
(iii) Lead	(c) Alloy of aluminium
(iv) Nickel	(d) Alloy of bronze

ANSWERS

A. True or False

1. False 2. True 3. False 4. True 5. True

B. Multiple Choice Questions

1. (c)	2. (b)	3. (c)	4. (b)	5. (c)
6. (a)	7. (c)	8. (c)	9. (b)	10. (a)
11. (a)	12. (b)	13. (c)	14. (b)	15. (c)
16. (a)	17. (b)	18. (b)	19. (c)	20. (a)
21. (b)	22. (d)			

C. Match the Following

1. (i)—(b), (ii)—(d), (iii)—(a), (iv)—(c)
2. (i)—(b), (ii)—(d), (iii)—(a), (iv)—(c)
3. (i)—(c), (ii)—(d), (iii)—(b), (iv)—(a)

❑❑❑

UNIT-III

Introduction to Metal Forming and Casting Process and its Application

5. METAL FORMING

6. CASTING

SYLLABUS

Metal Forming : Basic Metal Forming Operations and Uses of Such as : Forging, Rolling, Wire and Tubedrawing/Making and Extrusion, and Its Products/ Applications. Press-work Die and Punch Assembly, Cutting and Forming, its Applications. Hot-working Versus Cold-working.

Casting : Pattern and Allowances. Molding Sands and its Desirable Properties. Mould Making with the Use of a Core. Gating System. Casting Defects and Remedies. Cupola Furnace. Die-Casting and its Uses.

CHAPTER 5

Metal Forming

INSIDE THIS CHAPTER

5.1. INTRODUCTION

Forming is a method of shaping metal into desired form without any significant wastage of metal such as rolling, forging, and drawing. Metals are subjected to mechanical working for the following purposes :

1. To refine the grain structure.
2. To achieve high dimensional accuracy.
3. To impart good surface finish.
4. To improve the mechanical properties of the metal.

5.2. ELASTIC AND PLASTIC DEFORMATION

Deformation of metals is due to displacements of atoms and can be classified as :

5.2.1. Elastic Deformation

When force is applied on the metal within a limit, atoms move along the direction of force and occupy the new position, but on the removal of force, atoms come back to their original position. The limit upto which atoms come back to their original position is called elastic limit and this type of deformation is called elastic deformation.

5.2.2. Plastic Deformation

When force is applied on metal beyond the elastic limit, atoms doesn't come back to their original position on removal of applied force and occupy the new position. This type of deformation is called plastic deformation.

5.3. FORMING

Forming is defined as a process in which desired shape and size are obtained through the plastic deformation of the material under the action of external force. At elevated temperatures plastic deformation takes place in the material and new grains are formed at location of internal stresses by slip or twin deformation. This process of formation of new grains is called recrystallization and temperature at which new grains are formed is called recrystallization temperature.

There are two main categories of forming

1. Cold working
2. Hot working.

5.4. COLD WORKING [UPTU DEC. 2008]

It is defined as a process in which working temperature is lower than recrystallization temperature of the material. Cold working is performed at low temperature due to which oxidation and scaling of work material does not take place and there is less reduction of material. Cold working improves the hardness, yield strength, ultimate tensile strength and fatigue strength.

5.4.1. Advantages of Cold Working

(i) Good surface finish is obtained.
(ii) Strength, fatigue and wear properties are improved through strain hardening.
(iii) Directional properties can be achieved.
(iv) High dimensional control is achieved.
(v) Problem regarding contamination are less.
(vi) Suitable for mass production.

5.4.2. Disadvantages of Cold Working

(i) Heavy and powerful equipments are required.
(ii) High force is required to deform a metal plastically.
(iii) Undesirable and residual stresses are produced.
(iv) Annealing is required to compensate for the loss of ductility.
(v) Decrease in ductility and creep resistance of a metal.
(vi) Distortion in grain structure of a metal.

5.5. HOT WORKING

It is defined as a process in which working temperature is higher than recrystallization temperature of the material. Hot working refine the grain structure and improve the mechanical properties such as elongation, reduction of area, impact strength. During hot working defects in the metal such as blow holes, cracks, get welded up. It does not strain harden the metal.

5.5.1. Advantages of Hot Working

(i) Growth of grains takes place in range of working temperature so proper control of grain size is possible.

(ii) Due to absence of strain hardening sufficient amount of working can be imparted.

(iii) Elevated temperature reduces the strength and work load so large products and material of high strength can be worked under hot conditions by using low tonnage equipments.

(iv) Due to high ductility of material at high temperature, brittle material can be worked under hot working conditions.

5.5.2. Disadvantages of Hot Working

(i) Hot working is not used when very accurate dimensional control is required because of shrinkage and scaling.

(ii) There are some metals which are more brittle at high temperature so can not be hot worked.

(iii) At high temperature scaling of surface occurs and poor surface finish is obtained.

(iv) Dimensional tolerances are greater than that in cold worked metals.

(v) Structure and properties of metal are not uniform over the entire cross-section.

Various hot working processes are rolling, forging, extrusion, hot drawing, piercing, etc.

5.6. DIFFERENCE BETWEEN HOT WORKING AND COLD WORKING

[UPTU Dec. 2008, May 2009]

S.No	Hot Working	Cold Working
1.	Hot working is carried out at a temperature above recrystallization temperature but below the melting point.	Cold working is carried out at a temperature below the recrystallization temperature.
2.	Hot working refines metal grains which results in improvement of mechanical properties.	Cold working does not refine metal grains and lead to distortion of grains.
3.	Hardening of metal does not take place due to recovery and recrystallization.	Working below recrystallization temperature induces hardness in workpiece.

S.No	Hot Working	Cold Working
4.	During hot working internal or residual stresses are not developed.	During this process internal or residual stresses are developed.
5.	Stress required to cause deformation is less due to high deformation temperature.	Stress required to cause deformation is high due to low deformation temperature.
6.	Hot working does not affect hardness, yield strength, ultimate tensile strength and fatigue strength.	Cold working improves the hardness, yield strength, ultimate tensile strength and fatigue strength.
7.	Hot working improves mechanical properties such as elongation and impact strength.	Cold working reduces the impact strength and elongation.
8.	Hot working produces poor surface finish due to oxidation and scaling.	Good surface finish is obtained.
9.	It is difficult to maintain close dimensional tolerances.	High dimensional tolerances can be maintained.
10.	Existing cracks and blow holes are eliminated due to hot working.	In cold working, existing cracks propagate and new cracks may develop.

5.7. ROLLING [UPTU DEC. 2008, MAY 2009]

Rolling is defined as a process in which cross-section area of metal is reduced by passing the metal between circular or contoured rolls, rotating in opposite direction. Rolls rotate with surface velocity, which is higher than the speed of

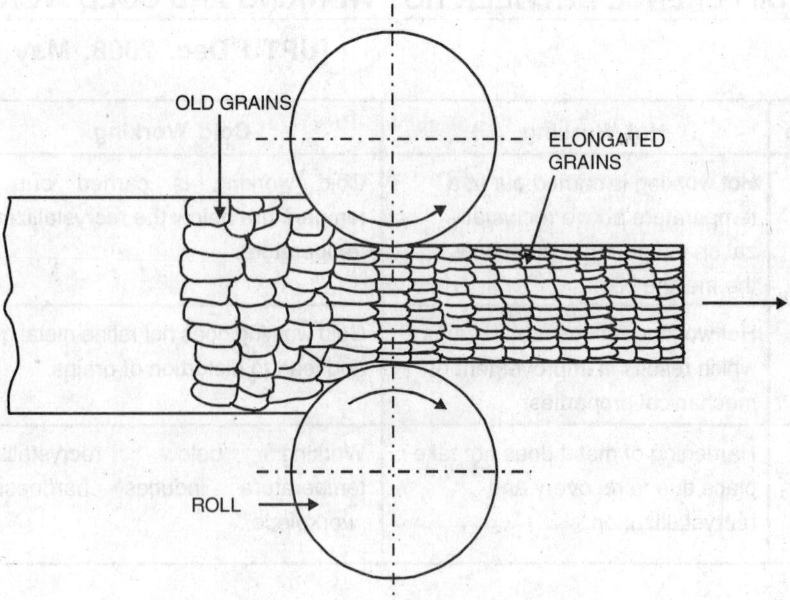

Fig. 5.1. Grain Reformation in Hot Rolling.

incoming metal. Due to excess speed, friction force acts between the rolls and metal, which try to propel the metal forward. The pressure on rolls builds up gradually from entry to neutral point (the point where pressure is highest) and then decreases till it reaches the exit. Due to pressure, metal is squeezed between the rolls and elongates to compensate for decrease in cross-sectional area.

This plastic deformation depends on friction conditions along the interface. If high friction is required, rolls cannot advance the material and skid over the surface.

5.7.1. Configuration of Rolling Mill

Arrangement of rolls in rolling mills are called rolling stand. Their names are based upon number of rolls employed. Most commonly used arrangements are :

(i) Two high non reversing rolling stand
(ii) Two high reversing rolling stand
(iii) Three high rolling stand
(iv) Four high rolling stand
(v) Cluster mill

(i) Two High Non-reversing Rolling Stand

It has two rolls positioned with one roll directly above the other. The rolls are supported in bearings and always move in only one direction. The upper roll may be raised or lowered to adjust the distance between the rolls.

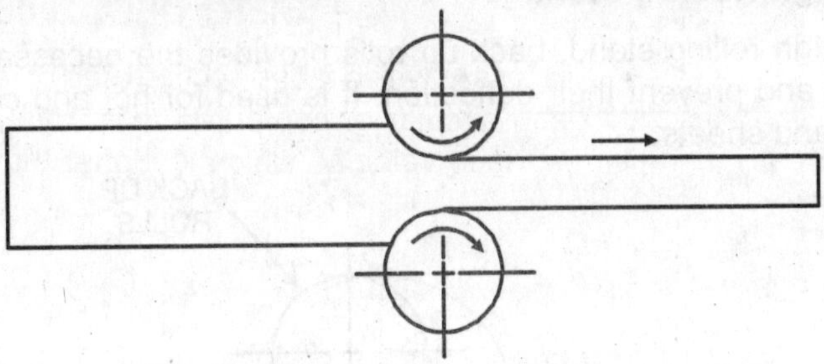

Fig. 5.2(*a*). Two High Non-Reversing Rolling Stand.

(ii) Two High Reversing Rolling Stand

In this rolling stand, direction of rotation of rolls can be reversed. This arrangement reduces the handling of hot metal in between the rolling passes but consumes more power as compared to non-reversing rolling stand.

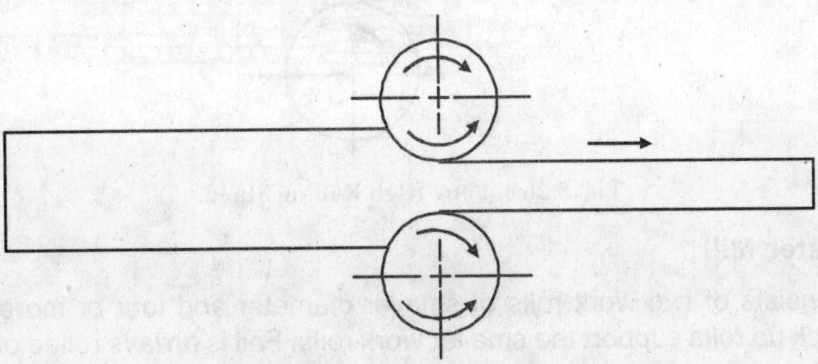

Fig. 5.2(*b*). Two High Reversing Rolling Stand.

(iii) Three High Rolling Stand

Three high rolling stand are used to eliminate the need of reversing the rolls. In this arrangement workpiece is feed in between upper and middle rolls in a direction opposite to the direction of feeding of workpiece between lower and middle rolls, resulting in high rate of production.

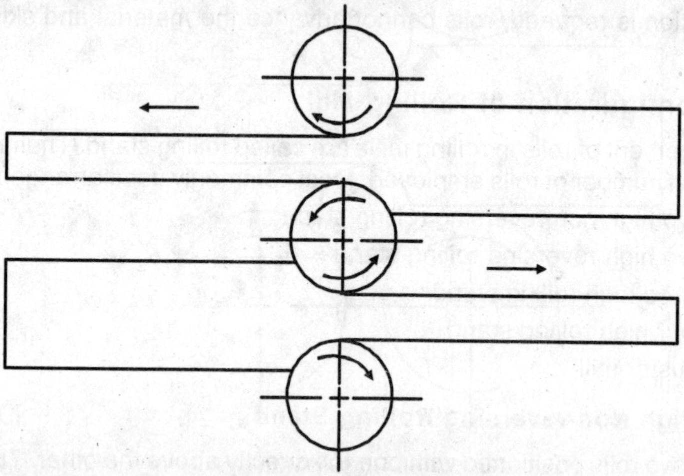

Fig. 5.2(*c*). Three High Rolling Stand.

(iv) Four High Rolling Stand

In four high rolling stand, back up rolls provides the necessary rigidity to smaller rolls and prevent their deflection. It is used for hot and cold rolling of wide plates and sheets.

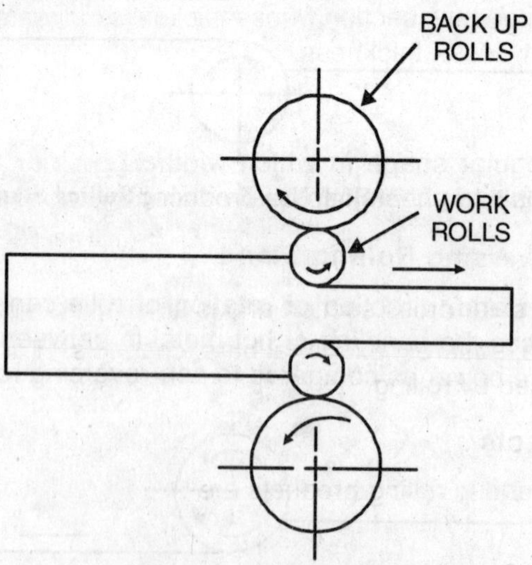

Fig. 5.2(*d*). Four High Rolling Stand.

(v) Cluster Mill

It consists of two work rolls of smaller diameter and four or more backup rolls. Back up rolls support the smaller work rolls. Foil is always rolled on cluster mills:

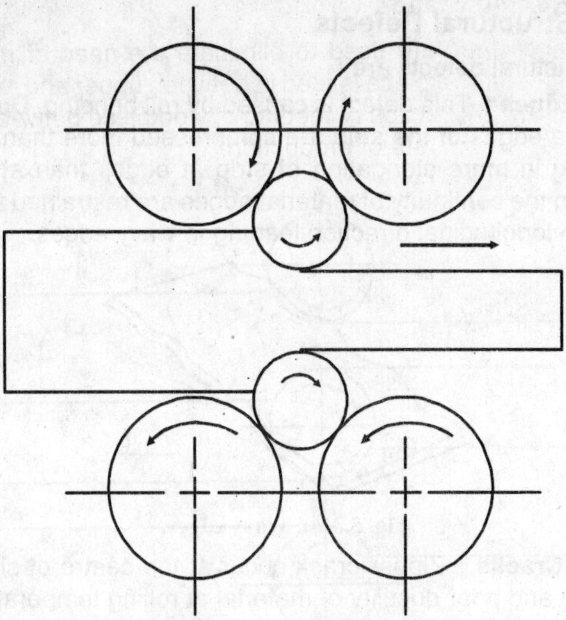

Fig. 5.2(e). Cluster Mill.

5.7.2. Rolling Products

Rolling is generally used to convert material into a finished product. There are many products such as blooms, billets, slabs which are obtained by rolling.

(i) Bloom

It has square or rectangular section whose thickness is greater than 6 inches and width is less than twice of thickness.

(ii) Slab

It also has rectangular shape in which width is greater than twice the thickness. Plates, strips and sheets can be produced by rolling slabs.

(iii) Billets

It has square or circular cross section and is smaller than bloom.

Sections like round, square, hexagonal bars, channels, H and I section like rail section are produced by rolling.

5.7.3. Rolling Defects

Various defects found in rolling products are :

(i) Surface Defects

Various surface defects are scale, cracks, scratches, pits, gauges and rust. These defects occur due to :

(a) Impurities and inclusions in the billet.

(b) Improper conditions related to material preparation and rolling operation.

(ii) Internal Structural Defects

Various structural defects are :

(a) **Wavy Edges :** This defect is caused by roll bending. Due to bending of rolls, the edges of the strip are compressed more than central portion resulting in more elongation of strip at edges than at centre. But to maintain the continuity of material edges are restrained from expanding freely in longitudinal direction leading to wavy edges.

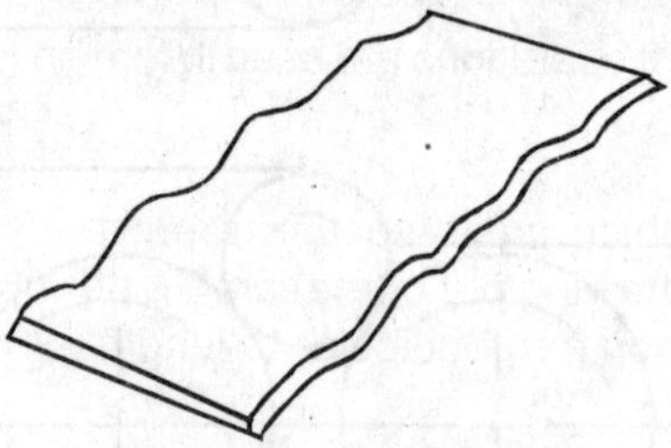

Fig. 5.3(*a*). Wavy edges.

(b) **Zipper Cracks :** Zipper crack occur at the centre of sheet due to roll bending and poor ductility of material at rolling temperature.

Fig. 5.3(*b*). Zipper cracks.

(c) **Alligatoring :** It is encountered due to non uniform deformation during rolling or due to presence of defect in billet. In alligatoring the sheet splits along a horizontal plane and follows the path of their respective rolls.

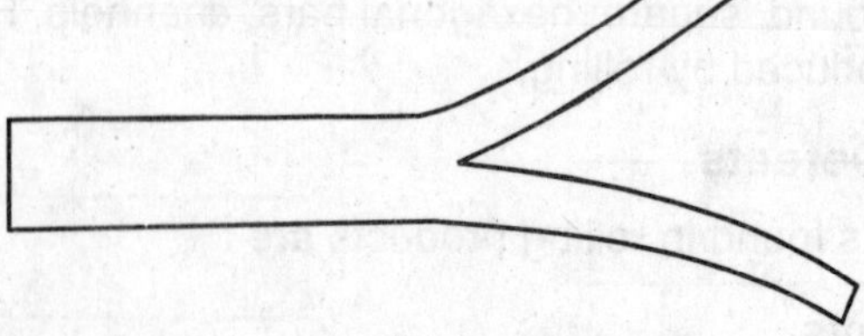

Fig. 5.3(*c*). Alligatoring.

(d) **Folds :** Folds occur during plate rolling due to very small reduction per pass.

(e) **Laminations :** Nonmetallic inclusions in material results in lamination along the thickness and results in reduction of strength of material along thickness direction.

5.8. FORGING [UPTU DEC. 2008, MAY 2009]

It is oldest metal working process. In ancient days forging was done by striking with a stone. But in modern forging strong arm hammer and anvil have been replaced by high powered hammers and mechanical presses.

In forging operation force is applied on heated metal manually or by means of power hammers, this applied force squeeze the metal between two or more dies and change the size and shape of the workpiece.

Generally two types of operations are used in forging to achieve the final object.

(a) Drawing Out

The operation in which cross-section area reduces and length of stock increases is known as drawing out operation. To perform this operation force is to be applied in a direction perpendicular to length axis.

(b) Upsetting

The operation in which cross-section area increases and length of stock reduces is known as upsetting operation. To perform the operation force is applied in a direction parallel to length axis.

5.8.1. Types of Forging

There are various types of forgings but some generally used are :

(i) Open die drop hammer forging
(ii) Impression die drop forging
(iii) Closed die forging
(iv) Upset forging
(v) Press forging
(vi) Roll forging
(vii) Swaging

(i) Open Die Drop Hammer Forging

In this forging the metal is heated to proper temperature and impact is delivered by mechanical hammer. It doesn't confine the flow of metal. The operator obtains the desired shape by orienting and positioning the workpiece between blows. It is used to induce plastic flow and minimize the machining.

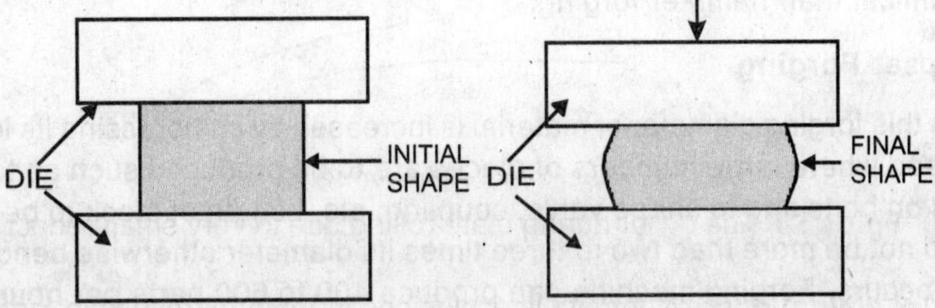

Fig. 5.4. Open Die Drop Hammer Forging.

(ii) Impression Die Drop Hammer Forging

It consists of a set of dies, one of which is attached to the hammer and other to anvil. Heated metal is positioned in lower die, which is strucked by upper die in one or more blows. Due to hammering metal flows and fill the die cavity completely. Excess metal, which is squeezed out around the periphery of the cavity is called flash.

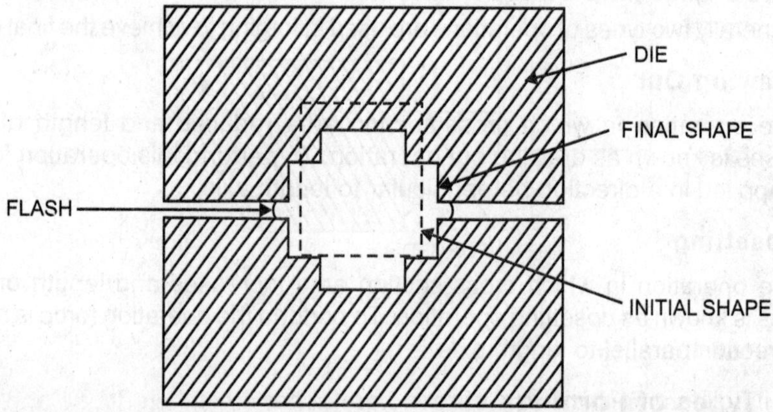

Fig. 5.5. Impression Die Drop Hammer Forging.

(iii) Closed Die Forging

Closed die forging is similar to impression die drop hammer forging but in this forging flash is not formed, so it is called flashless forging. Workpiece completely fill the die cavity without any loss of material so accurate workpiece sizing is required. To obtain forging of desired dimension and tolerance, proper die design are required. Size of blank must not be over, otherwise it can generate excessive pressure which can damage or jam the die. In case of undersized blank, die cavity is not completely filled up.

(iv) Press Forging

In press forging metal is shaped by single continuous action instead of a series of blows as in drop forging. A great proportion of the workdone is transmitted to the workpiece as compared to hammer forging in which much of the work is absorbed by machine and foundation. Press forging is more economical than hammer forging.

(v) Upset Forging

In this forging diameter of material is increased by compressing its length. It is used where large numbers of pieces are to be produced such as to form heads on bolts and to shape valve, coupling, etc. Length of stock to be upset should not be more than two to three times its diameter otherwise bending of stock occurs. Forging machine can produce 400 to 600 parts per hour.

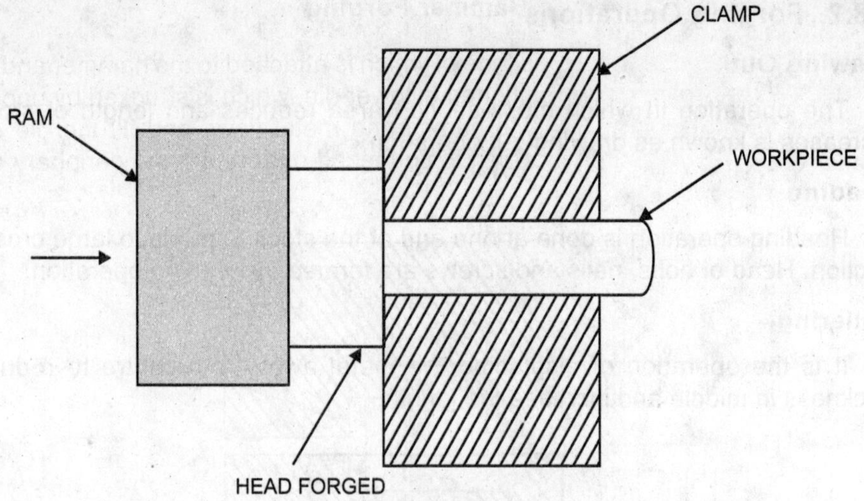

Fig. 5.6. Upset Forging.

(vi) Roll Forging

In roll forging thickness of round or flat bar stock is reduced and length is increased. Forging machine have two cylindrical or semi-cylindrical rolls, containing one or more grooves to impart a desired shape to the workpiece. Main applications of roll forging are found in leaf spring, axles, levers, etc.

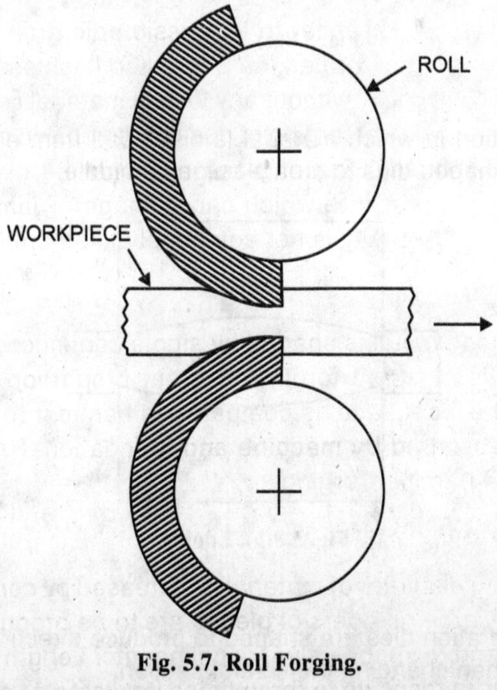

Fig. 5.7. Roll Forging.

(vii) Swaging

It is used to reduce the diameter of rod or tube to impart a desired shape to the piece. In swaging die itself act as a hammer and metal is hammered from various angles due to which metal flows inward and take the shape of die.

5.8.2. Forging Operations

Drawing Out

The operation in which cross-section area reduces and length of stock increases is known as drawing out operation.

Heading

Heading operation is done at one end of the stock to produce large cross-section. Head of bolts, nails and screws are formed by heading operation.

Fullering

It is the operation of displacing the metal away from centre to reduce thickness in middle and increase the length.

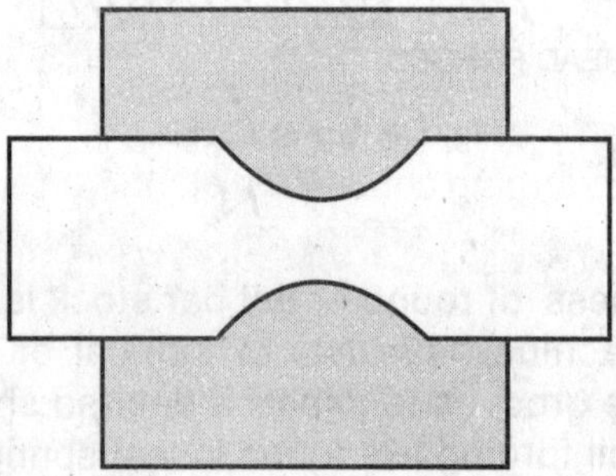

Fig. 5.8(*a*). Fullering.

Edging

It is the operation in which material is displaced from edges to inside i.e. between two cup shaped dies to create bulge in middle.

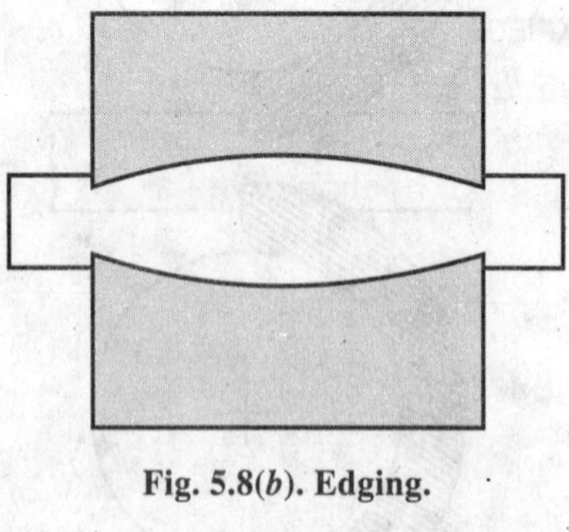

Fig. 5.8(*b*). Edging.

Bending

In bending operation dies are shaped to produce directional change in the workpiece rather than changing the size of section.

Upsetting

The operation in which cross-section area increases and length of stock reduces is known as upsetting operation.

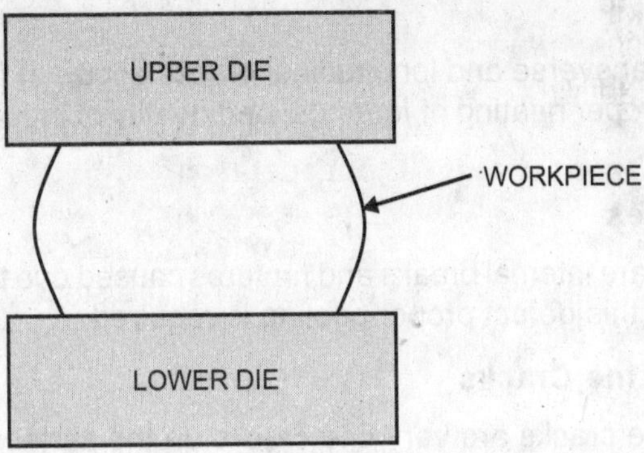

Fig. 5.8(*c*). Upsetting.

5.8.3. Forging Defects

Main causes due to which defect occur in forgings are improper heating, incorrect die design, uneven cooling of stock, defect in the original metal, improper forging operation, incorrect forging conditions, etc. Some common defects found in forgings are :

(i) Cold Shuts

Cold shuts are short cracks. They are caused due to folding of metal surface against itself during forging. It can be avoided by rounding the sharp corners.

(ii) Pitting

Pitting is caused by scales which is not completely removed from the die cavities and is worked into the surface of the forging. Scale produces shallow depressions on the surface of forging. It can be avoided by cleaning of dies and proper control of forging temperature.

(iii) Incomplete Filling of Dies

This defect occurs due to incomplete filling of closed impression die cavity. It occurs due to many reasons such as poor forging design, wrong amount of stock, incorrect die design, poor heating of metal, insufficient number of blows during forging, etc.

(iv) Die Shift

Die shift is caused due to upper and lower die misalignment. This defect is caused due to loose wedges. So wedges should be rigidly fixed.

(v) Fins and Rags

Fins and rags are small projections or pieces of loose metal driven into the surface of forging. To control this defect proper die design is required.

(vi) Burnt and Overheated Metal

This defects occurs when the metal is heated for too long time or at too high temperature. Proper heating is required to control this defect.

(vii) Cracks

Both transverse and longitudinal cracks occur on forging surface. It occurs due to improper heating of forgings, bad quality of ingot and improper cooling of alloy steels.

(viii) Flakes

Flakes are internal breaks and ruptures caused due to rapid cooling of forging. To remove this defect proper cooling is required.

(ix) Hair Line Cracks

Hair line cracks are very fine cracks on the surface of the forgings caused due to rapid cooling of forging and defective metal piece.

(x) Dents

Dents are produced due to careless working.

5.9. EXTRUSION [UPTU DEC. 2008, MAY 2009]

It is one of the most useful metal forming process and may be performed in hot or cold conditions. Hot extrusion is commonly used for metals to reduce the required forces but it has number of problems such as die wear and cooling arrangement.

In extrusion metal filled in closed cavity is forced to flow through small opening to form a product which take the shape of opening and is similar to squeezing of toothpaste out of toothpaste tube.

For extrusion hot metal billet is filled in chamber and die plate with necessary opening is fixed in position at one end of container. As the ram advances it compress the metal and force it to flow through die opening. Dummy block is made of steel and kept between hot billet and ram to protect ram from heat and pressure. In extrusion no draft is required and dies are relatively inexpensive. Extrusion has a limitation that it can only produce same and uniform cross-section for the entire length.

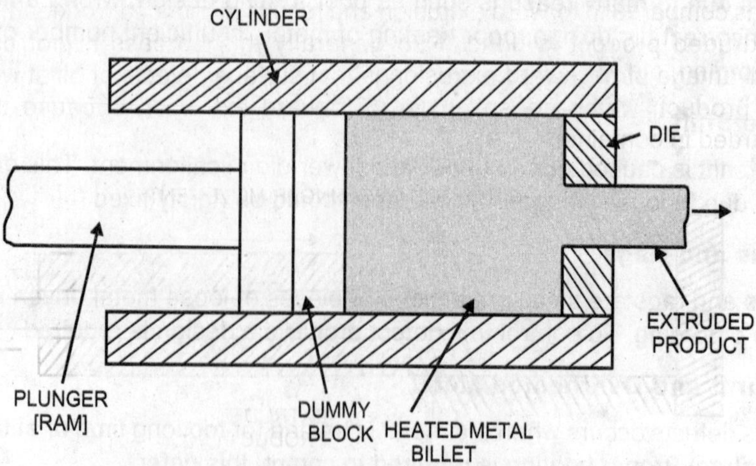

Fig. 5.9. Extrusion Process.

Commonly used metals for extrusion are lead, magnesium, aluminum due to low yield strength and low hot working temperature. Steel, stainless steels are difficult to extrude due to high yield strength and tendency to weld to the walls of die. Extrusion process is used to manufacture railings, I-Beams, T-section beams, channels, pipes, hollow bars, etc.

5.9.1. Extrusion Process

(i) Hot Extrusion Process

For hot extrusion, generally two type of process are used :

(a) Forward Extrusion or Direct Extrusion : In this extrusion the direction of motion of ram and extruded product are same. It means metal flows in forward direction. In this process problem of friction exist due to relative motion between heated metal billet and cylinder walls. Due to large frictional loss and high working load, container is subjected to high radial stress. Lubricants are used to reduce the friction.

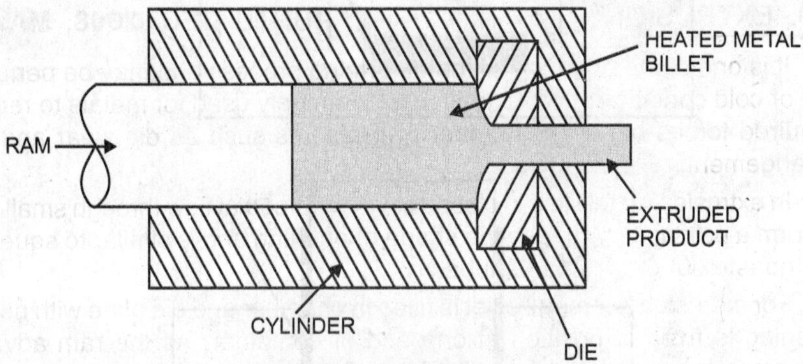

Fig. 5.10(*a*). Direct Extrusion.

(b) Backward or Indirect Extrusion : To avoid the problem of friction we use backward extrusion in which billet remains stationary and product is extruded in opposite direction of ram movement. Friction force acts only at die container interface and is absent in between the billet and container. Thus friction is very less as compared to forward extrusion and results in reduced workload. Handling of extruded product is difficult so generally this process is not used. The disadvantage of backward extrusion is that surface defects of billet will exist in final product, which doesn't exist in forward extrusion because these are discarded in container.

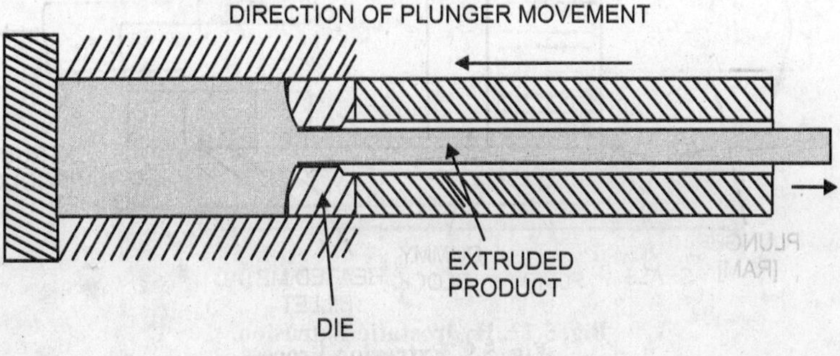

Fig. 5.10(*b*). Indirect Extrusion.

(ii) Cold Extrusion Process

(a) Impact Extrusion : This process is used for soft and ductile materials and operation is performed under cold condition. To form a component slug is kept on die and punch strikes the slug against the die. Due to application of pressure, metal extrudes through the gap between punch and die opposite to the punch movement and slides along the sidewall. Height of sidewalls is controlled by amount of metal in slug. Thin walled cans, collapsible tubes are formed by impact extrusion.

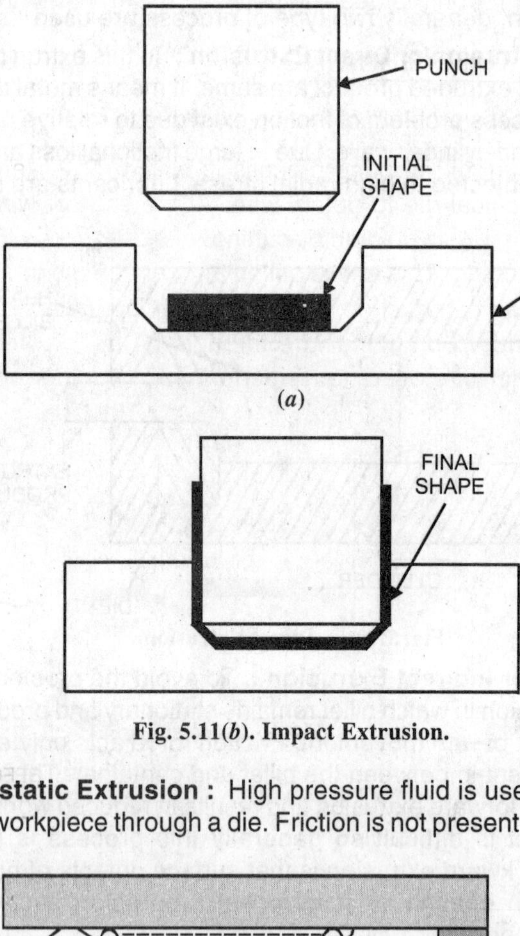

Fig. 5.11(*b*). Impact Extrusion.

(b) Hydrostatic Extrusion : High pressure fluid is used to apply the force to extrude the workpiece through a die. Friction is not present in the billet chamber

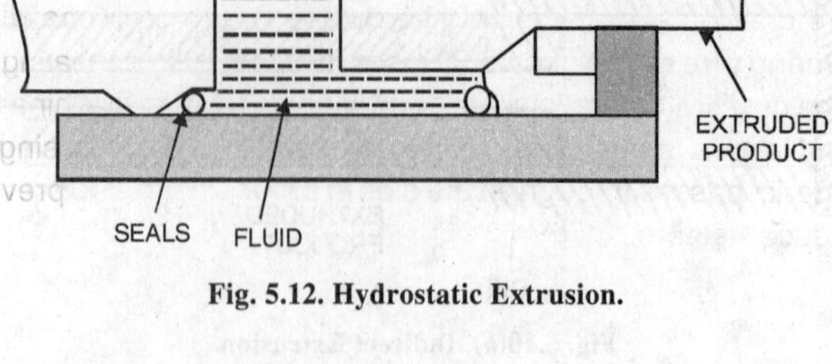

Fig. 5.12. Hydrostatic Extrusion.

due to use of fluid, which act as lubricant. By slight variation in the process it also provides facility of extruding a relatively brittle material. At high temperature common fluid decomposes so temperature range is limited. Leak proof seal must be designed to contain the pressurized fluid without leakage.

5.9.2. Extrusion of Hollow Shapes

Mandrel is used along with the ram to form the bore of tube. Mandrel may be either be fixed to ram or frame. Plunger applies the pressure on billet and metal is forced out through the opening between die and mandrel.

5.10. WIRE AND TUBE DRAWING [UPTU DEC. 2008]

5.10.1. Wire Drawing

Wire drawing is a cold working process in which rod of large diameter is passed through a conical die to obtain wire. At the start of wire drawing, the front end of the rod or wire is pointed by cutting or squeezing so that it can freely enter through die opening. This end is pulled through the die to obtain sufficient length so that wire can be attached in jaws of power operated reel. Reel is then rotated at desired rate to draw the wire through the die. To reduce the large diameter into very thin wire operation is performed in number of passes.

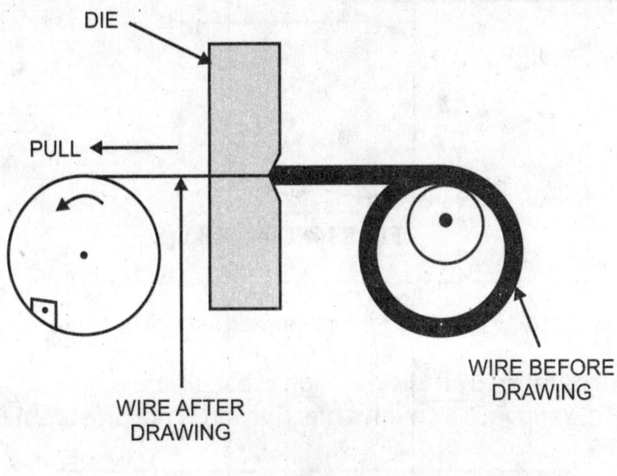

Fig. 5.13.

Contact region of die are made of tungsten carbide or diamond. Wire drawing speed for large diameter is 10 m/min and for very thin wire is 1800 m/min. In this process sufficient amount of heat is generated which is removed by water.

For wire drawing material should be ductile otherwise proper annealing is required. During wire drawing material loses its ductility so annealing is required to restore the ductility. Operations are performed on tandom machine to minimize handling and labour. In this system 3 to 12 dies are mounted in a single machine and the material passes through all the dies in proper manner to prevent tension that can induce fracture.

5.10.2. Tube Drawing

Tube drawing is cold working process which is used to draw tubes or pipes by using hollow seamless and weided tube shells produced through hot working. These tube shells are cold drawn to produce tubes with smooth surface, thin walls and better dimensional accuracy. Operation is started by swaging the diameter of starting end to a smaller diameter to enable to pass through die. After passing through the die, end is gripped in tongs and connected to draw bench for tube drawing. Outer diameter is controlled by die and internal diameter is controlled by internal mandrel. If mandrel is not used in tube drawing, then it is known as tube sinking.

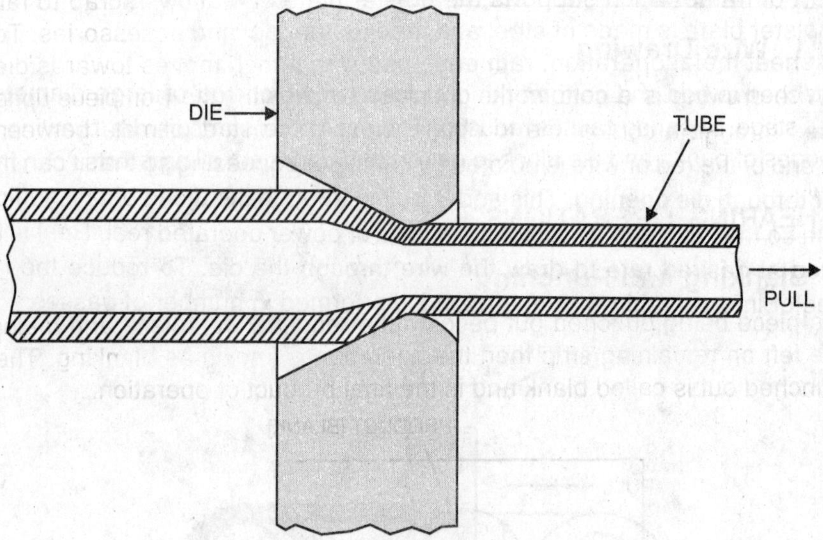

Fig. 5.14. Tube Sinking.

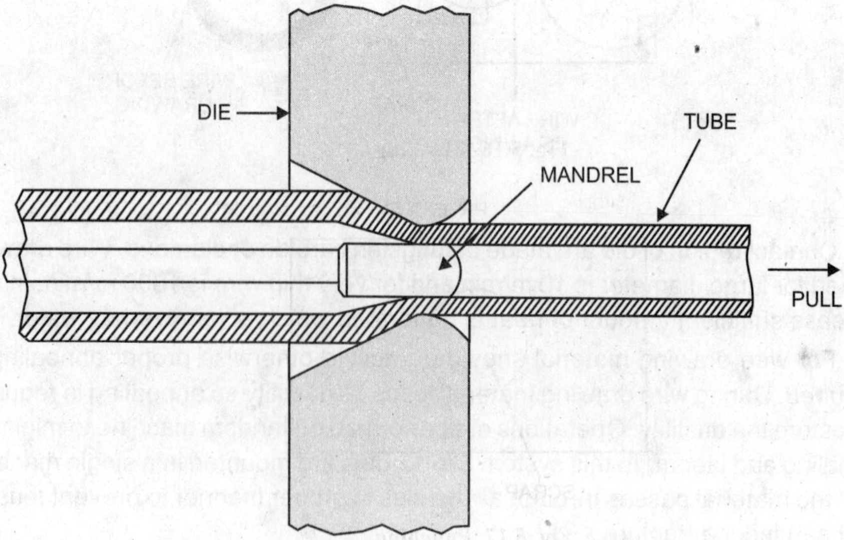

Fig. 5.15. Tube Drawing with Mandrel.

5.11. PRESS WORK

Press working or sheet metal working process is a manufacturing process in which parts are produced without formation of chip. Press working is most widely used method due to low cost, good surface finish and exact shape as compared to forging and casting. It also produces large number of components in small time. It is used for producing car bodies, beverage cans, cabinets, household utensils, etc. In sheet metal working, product obtained from forming operations such as plates, bars, tubes, sheets, etc. are processed in mechanically or hydraulically operated press.

Major components of press are bed, ram, bolster plate, cushion etc. Bed is the part of frame which supports the bolster plate and allows scrap to fall down. Bolster plate is made of steel and mounts the tool and accessories. To perform sheet metal operation, ram equipped with punch moves towards die block attached to bed and perform the operation by shearing the workpiece up to the plastic stage. Shearing causes reduction of cross-section area of metal between cutting edges of punch and die block to generate fracture at cutting points.

5.12. SHEARING OPERATIONS

5.12.1. Blanking and Punching

If the piece being punched out become the required workpiece and major burrs are left on remaining strip then the operation is known as blanking. The piece punched out is called blank and is the final product of operation.

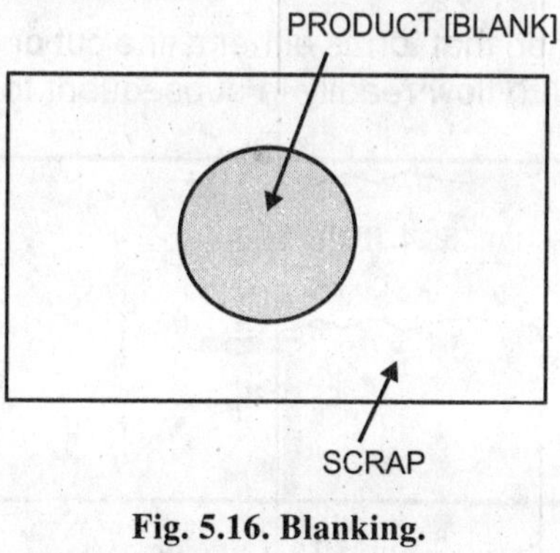

Fig. 5.16. Blanking.

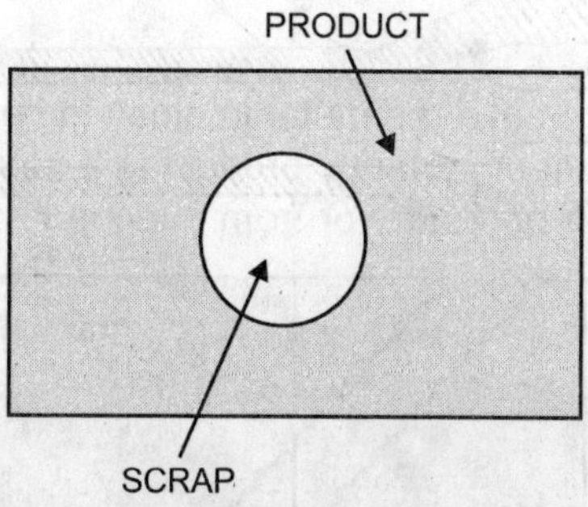

Fig. 5.17. Punching.

If punch out is scrap and remaining strip is required product then the operation is known as punching. Punching is used to form various shaped holes in sheet metal.

5.12.2. Notching

Notching is similar to punching except that edge of strip forms part of the punch out perimeter. In this operation a specified portion of metal towards the edge of stock is removed.

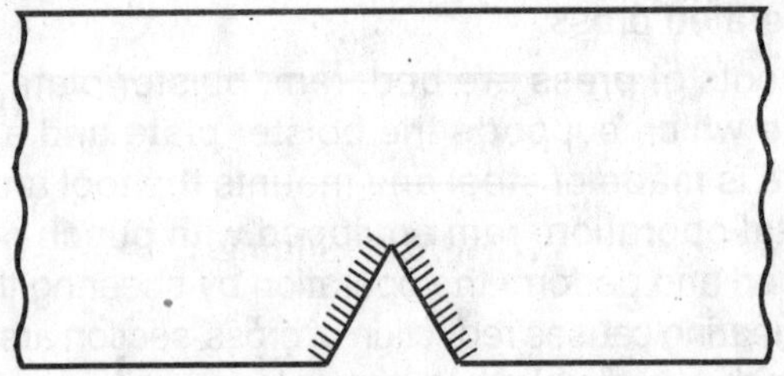

Fig. 5.18. Notching.

5.12.3. Trimming

In trimming operation small amount of unwanted material known as flash is removed from parting plane. Flash is generally formed during die casting and drop forging.

5.12.4. Lancing

It is a cutting operation that forms either a line cut or an actual hole in metal. It allows adjacent metal to flow readily in subsequent forming operation.

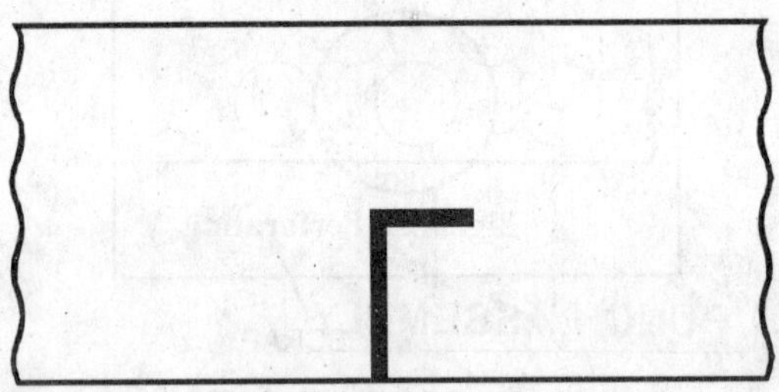

Fig. 5.19. Lancing.

5.12.5. Nibbling

Nibbling is used to remove the metal in small increments. In this operation contour is cut out from sheet metal by producing a series of overlapping slit or notches either from outer boundary or from punched hole.

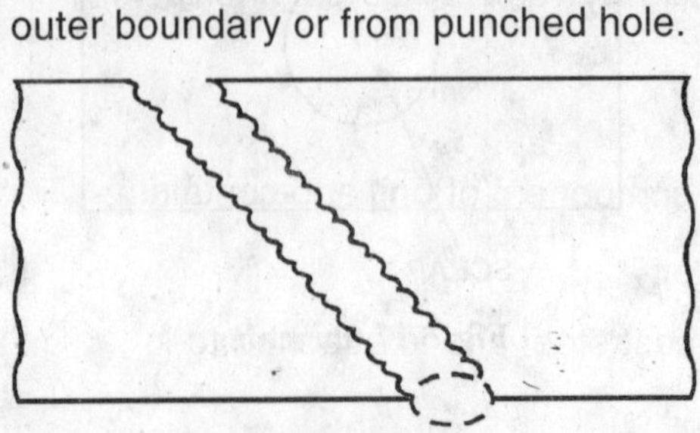

Fig. 5.20. Nibbling.

5.12.6. Slitting

It is used to produce narrow strips in which a wide coiled sheet metal is cut lengthwise. These strips have accurate and constant width because distance between adjacent set of shearing edges is fixed.

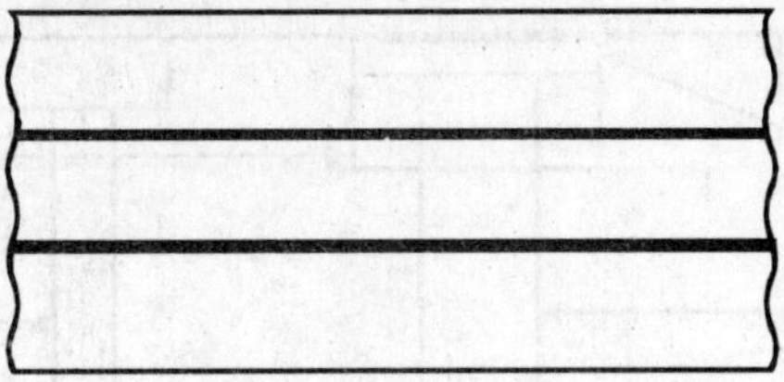

Fig. 5.21. Slitting.

5.12.7. Shaving

It is a finishing operation in which small amount of metal is removed from the edge of already blanked part. Punches and dies are made with very small clearance to remove small amount of metal.

5.12.8. Perforating

In this operation large numbers of closely spaced holes are produced. These holes are identical and arranged in regular pattern.

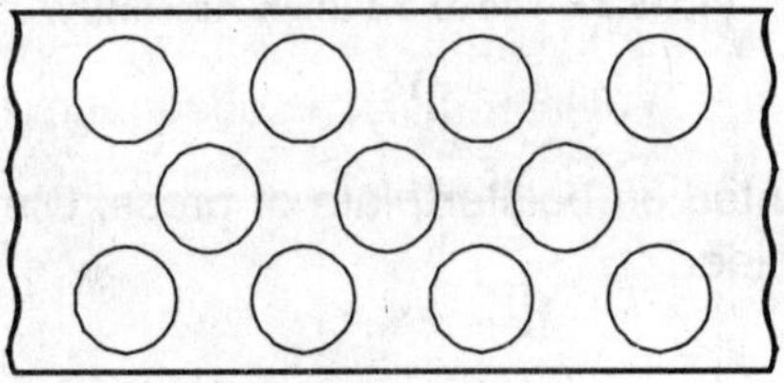

Fig. 5.22. Perforating.

5.13. DIE AND PUNCH ASSEMBLY

Die set consist of punch, die and other components known as accessories. Details about the accessories are described as follows

(1) Punch

It is the male component of unit which is made up of hard and wear resistant metal and is moved by ram. It has small size having sufficient strength and rigidity.

(2) Die

It is a female component of unit and contains a die cavity.

(3) Punch Holder

It is made of cast steel and is attached to lower end of press ram with the help of shank.

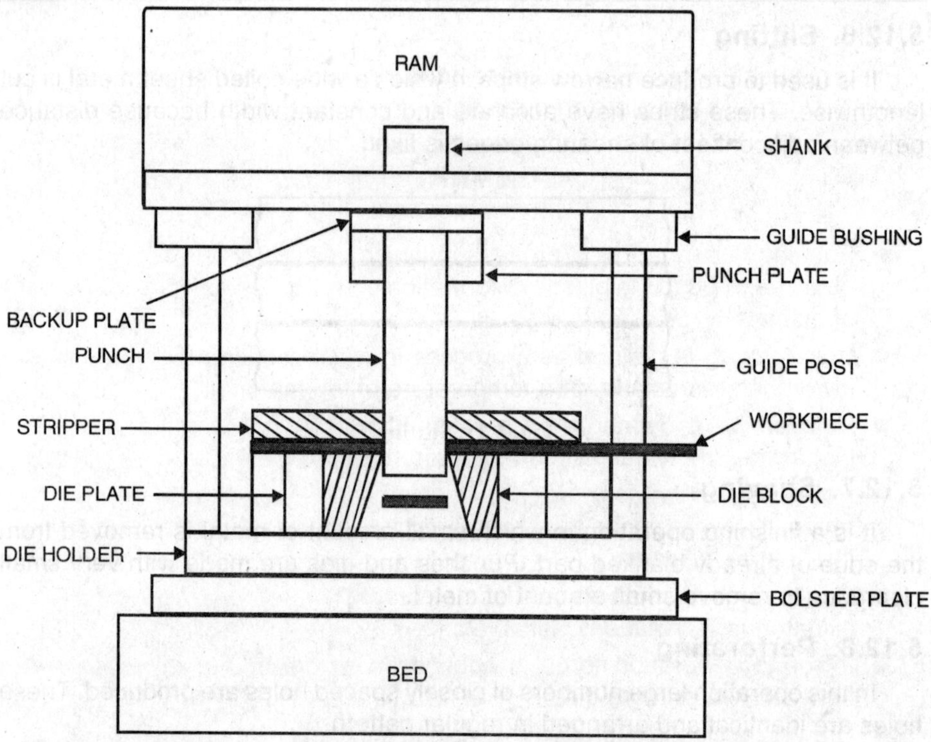

Fig. 5.23. Die and Punch Assembly.

(4) Die Holder

Die holder is mounted on bolster plate of press. Die block and guide post are mounted on die holder.

(5) Punch Plate

It fits over the punch body and holds it in proper position and it is also called as punch retainer.

(6) Bolster Plate

It is thick plate, which is used for locating and supporting die assembly.

(7) Back up Plate

It is also called pressure plate and used to reduce excessive pressure on punch plate.

(8) Guide Post and Bushings

Guide post is used to hold the punch and die in proper alignment during operation.

(9) Stripper Plate

It is a plate which removes the scrap from punch or die and also guide the metal sheet.

(10) Pilots

Pilots are used in progressive dies to bring the stock into correct position for successive blanking and piercing operation.

SUMMARY

- Forming is defined as a process in which desired shape and size are obtained through the plastic deformation of the material under the action of external force.
- Cold forming is defined as a process in which working temperature is lower than recrystallization temperature of the material.
- Hot forming is defined as a process in which working temperature is higher than recrystallization temperature of the material.
- Rolling is defined as a process in which cross-section area of metal is reduced by passing the metal between circular or contoured rolls, rotating in opposite direction.
- Various rolling defects are scale, cracks, scratches, wavy edges, alligatoring, laminations, folds, etc.
- In forging operation force is applied on heated metal to squeeze it between two or more dies.
- Drawing out is the operation of reducing the cross-section and increasing the length whereas upsetting is the operation of increasing the cross-section and decreasing the length.
- In extrusion, metal filled in closed cavity is forced to flow through small opening to form a product which takes the shape of opening.
- In forward extrusion direction of motion of ram and extruded product are same whereas in backward extrusion product is extruded in opposite direction of ram movement.
- Wire drawing is a cold working process in which rod of large diameter is passed through a conical die to obtain wire.
- Tube drawing is cold working process which is used to draw tubes or pipes by using hollow seamless and welded tube shells produced through hot working.
- Press working or sheet metal working process is a manufacturing process in which parts are produced without formation of chip.
- If the piece being punched out become the required workpiece and major burrs are left on remaining strip then the operation is known as blanking.
- If punch out is scrap and remaining strip is required product then the operation is known as punching.

SHORT QUESTIONS WITH ANSWERS

Q. 1. What is the difference between elastic and plastic deformation?

Ans. In elastic deformation atoms come back to their original position on removal of applied force but in plastic deformation atoms occupy new position and they don't come back to their original position.

Q. 2. What is recrystallization temperature?

Ans. The temperature at which grain reformation takes place is known as recrystallization temperature.

Q. 3. Give two examples of hot working and cold working process.

Ans. Hot working : Rolling, Extrusion.

Cold working : Wire drawing, Sheet metal process.

Q. 4. Why the brittle material can be worked under hot working conditions?

Ans. Generally at high temperature ductility of material increases so brittle material can be worked under hot working condition.

Q. 5. What is use of back up rolls in rolling mills?

Ans. Back up rolls support the small work rolls and provide the necessary rigidity to small rolls to prevent their deflection.

Q. 6. Give some examples of rolling products.

Ans. Bloom, Slab, Billets.

Q. 7. Why the wavy edges are formed on surface of workpiece during rolling?

Ans. Wavy edges are formed due to roll bending in which workpiece is thicker at centre than at edges.

Q. 8. What are two operations which are required in forging to obtain final object?

Ans. (1) Drawing out (2) Upsetting

Q. 9. What is the difference between impression die drop forging and closed die forging?

Ans. In impression die drop forging flash is formed but in closed forging flash is not formed.

Q. 10. Why oversized blank is not used in closed die forging?

Ans. Oversized blank can generate excessive pressure which can damage or jam the die.

Q. 11. Give any two examples of forging defects.

Ans. (1) Cold shut (2) Flakes

Q.12. What is the difference between direct extrusion and indirect extrusion?

Ans. In direct extrusion the direction of motion of ram and extruded product are same but in indirect extrusion direction of motion of ram is opposite to extruded product.

Q. 13. What is the use of mandrel in extrusion?

Ans. Mandrel is used to form the hollow shapes.

Q. 14. What is the use of trimming operation?

Ans. Trimming operation is used to remove the unwanted material.

Q.15. What is use of stripper plate?

Ans. Stripper plate is used to remove the scrap from punch or die.

REVIEW QUESTIONS

1. What do you mean by forming operations? Explain in brief with examples and neat sketches.
2. Explain the advantages and disadvantages of hot and cold working.
3. Define recrystallization temperature and its significance in metal forming.
4. What do you mean by rolling? Explain the principle with neat sketch and its applications. **[UPTU Dec. 2009]**
5. What do you mean by forging? Explain different type of forging with neat sketches.
6. Define extrusion with example. Also explain hydrostatic extrusion with sketch.

7. Write short notes on
   (a) Direct extrusion
   (b) Indirect extrusion
   (c) Impact extrusion
8. Explain the forging and rolling defects.
9. Write the difference between wire and tube drawing.
10. Give applications of forging, extrusion and rolling process.
11. Compare hot working and cold working processes.
12. With the help of sketch, define the following operations
   (a) Forging (b) Rolling
   (c) Drawing (d) Extrusion **[UPTU Dec. 2008]**
13. Differentiate between hot and cold working of metals. Bring out the advantages and disadvantages of each of these technique. Explain with neat sketch rolling of steel bar. **[UPTU Dec. 2008, 2009]**
14. List the advantage of forging of metals. Why is press forging preferred over hammer forging process. **[UPTU Dec 2011]**
15. Draw a neat sketch of die punch assembly with acessory parts or components. Explain how cutting operations such as blanking or piercing differ from deep drawing operations. **[UPTU May 2012]**

SELF TEST QUESTIONS

A. State Whether True or False

1. In cold working, working temperature is higher than recrystallization temperature of material.
2. Cold working provides good surface finish.
3. Large amount of strain hardening makes the material ductile.
4. Forging is done by means of high powered hammers and mechanical presses.
5. In upsetting operation, cross-section area of work piece reduces.
6. Undersize blank in closed die forging may jam the die.
7. Dents formation takes place due to improper working.
8. In indirect extrusion the direction of movement of ram and extruded product is same.
9. Impact extrusion is cold working process.
10. If punch-out is scrap and remaining strip is required product, then the operation is known as blanking.
11. Misrun is a metal forming defect. **[UPTU Dec. 2009]**

B. Fill in The Blanks

1. Formation of new grains, due to plastic deformation is called..............
2. In ...., working temperature is higher than recrystallization temperature of material.
3. .... provide the necessary rigidity to smaller rolls and prevent their deflection.
4. Nonmetallic inclusions in material result .... along the thickness.
5. ....is also called flashless forging.

6. .... are internal breaks occurring due to rapid cooling from forging temperature.
7. .... produces tubes from cold working of seamless and welded tube shells.
8. In .... processes parts are produced without formation of chip.
9. .... is a female working member of die and punch assembly which contain a die cavity.
10. .... plate is used for locating and supporting the assembly.

C. Multiple Choice Questions

1. Which arrangement is used to form foils
   (a) Two high rolling mill
   (b) Three high rolling mill
   (c) Four high rolling mill
   (d) Cluster mill
2. The defect that occur due to roll bending in which resulting strip is thicker at centre than at edge.
   (a) Folds (b) Alligatoring
   (c) Wavy edges (d) Laminations
3. Which forging is used to form bolts?
   (a) Upset Forging (b) Press Forging
   (c) Roll Forging (d) Swaging
4. Defect due to misalignment of die halves.
   (a) Cold shut (b) Flakes
   (c) Die shift (d) Scale pits
5. Process which produces the same and uniform cross-section for the entire length.
   (a) Extrusion (b) Coining
   (c) Forging (d) Blanking
6. Which part is used in extrusion process to form hollow tubes?
   (a) Ram (b) Plunger
   (c) Mandrel (d) None
7. Wire drawing speed for large diameter wire is
   (a) 2 m/min (b) 10 m/min
   (c) 50 m/min (d) 100 m/min
8. Which process causes reduction of cross-section area of metal between cutting edges of punch and die block to generate fracture at cutting points.
   (a) Extrusion (b) Forging
   (c) Shearing (d) None
9. Operation in which large numbers of closely spaced holes are produced.
   (a) Shaving (b) Perforating
   (c) Trimming (d) Lancing
10. Finishing operation in which burrs are removed from the edge of blank part.
    (a) Shaving (b) Perforating
    (c) Trimming (d) Lancing
11. Forging of metal is a **[UPTU Dec. 2008]**
    (a) Machining process (b) Welding process
    (c) Joining process (d) Metal forming process

12. Metal forming is the process in which [UPTU Dec. 2008]
    (a) Removal of metal takes place
    (b) Shaping of metal by forces applied through various dies and tools.
    (c) Some additional materials are joined to raw material
    (d) Material addition, through melting and pouring
13. The limit upto which atoms come back to their original position on removal of external force is called ________
    (a) Plastic limit (b) Yield limit
    (c) Elastic limit (d) None of these
14. The process in which desired shape and size are obtained through plastic deformation of material under the action of external force
    (a) Forming (b) Casting
    (c) Machining (d) Welding
15. Recrystallization is
    (a) Distortion of new grains (b) Formation of new grains
    (c) Breaking of new grains (d) None of these
16. The process in which working temperature is higher than recrystallization temperature
    (a) Cold working (b) Hot working
    (c) Cold rolling (d) All of these
17. Which forming process is used to form thin walled cans, collapsible tubes
    (a) Finish rolling (b) Impact extrusion
    (c) Press working (d) Forging
18. Disadvantage of cold working is
    (a) Residual stress is produced
    (b) High dimensional control is achieved
    (c) Suitable for mass production
    (d) Directional properties are achieved
19. Which process is required to compensate for loss of ductility
    (a) Cyaniding (b) Carburizing
    (c) Case hardening (d) Annealing
20. Mechanical property which is not improved by hot working
    (a) Elongation (b) Impact strength
    (c) Reduction of area (d) Hardness
21. Function of back up rolls is to
    (a) Provide flexibility to smaller rolls
    (b) Provide rigidity to smaller rolls
    (c) Deflect the rolls
    (d) All of these
22. Which mill is used to eliminate the need of reversing the rolls
    (a) Two high reversing rolling stand
    (b) Three high rolling stand
    (c) Four high rolling stand
    (d) Cluster mill
23. The rolling product is
    (a) Bloom (b) Slab
    (c) Billets (d) All of these

24. Which is not the surface defect

(a) Pits (b) Folds

(c) Gauges (d) Scratches

25. Defect caused due to roll bending is

(a) Laminations (b) Alligatoring

(c) Wavy edges (d) Folds

26. Lamination occurs due to

(a) Roll bending (b) Nonmetallic inclusion

(c) Small reduction per pass (d) All of these

27. Process in which force is applied by means of hammer

(a) Casting (b) Forging

(c) Extrusion (d) Rolling

28. The operation in which cross-section area increases and length of stock decreases

(a) Drawing (b) Swaging

(c) Rolling (d) Upsetting

29. Which is not the type of forging

(a) Blow forging (b) Upset forging

(c) Press forging (d) Roll forging

30. The operation in which force is applied perpendicular to length axis

(a) Drawing (b) Upsetting

(c) Perforating (d) Notching

31. During forging, excess metal which is squeezed out around the periphery of cavity is called ________

(a) Trim (b) Flash

(c) Blank (d) Stock

32. The forging in which flash is not formed is called

(a) Open die drop hammer forging

(b) Impression die hammer forging

(c) Closed die forging

(d) All of above

33. If size of blank to be forged is over in closed die forging then it can

(a) Melt the die cavity

(b) Damage the blank

(c) Damage or jam the die (d) All of these

34. Which forging is used to form head on bolts

(a) Roll forging (b) Closed die forging

(c) Swaging (d) Upsetting

35. The operation in which thickness of round or flat bar stock is reduced and length is increased by use of semicylindrical rolls.

(a) Swaging (b) Roll forging

(c) Upsetting (d) Drawing

36. The operation in which thickness is reduced in middle by displacing the metal away from centre

(a) Heading (b) Edging

(c) Fullering (d) Upsetting

37. Defects in forging occurs due to
(a) Improper heating (b) Uneven cooling of stock
(c) Incorrect die design (d) All of these

38. Cold shut can be avoided
(a) By cooling
(b) By heating
(c) By rounding the sharp corners
(d) By cleaning of die

39. Forging defect which occur due to upper and lower die misalignment is called as
(a) Die shift (b) Cold shut
(c) Pitting (d) Flakes

40. Hair line cracks occur in the surface of forging due to
(a) Careless working (b) Rapid cooling
(c) Poor heating (d) Slow cooling

41. The defect caused due to uncomplete removal of scales is
(a) Cold shuts (b) Die shift
(c) Pitting (d) Dents

42. The forming process which is similar to squeezing of toothpaste out of toothpaste tube
(a) Forging (b) Rolling
(c) Extrusion (d) Press working

43. The process in which metal filled in closed cavity is forced to flow through small opening to form a product
(a) Extrusion (b) Forging
(c) Rolling (d) Press working

44. A piece made of steel, kept between hot billet and ram to protect ram from heat and pressure is called
(a) Punch (b) Dummy block
(c) Die (d) Ingot

45. Stainless steel are difficult to extrude due to
(a) High yield strength
(b) Low yield strength
(c) Low hot working temperature
(d) None of these

46. Commonly used metal for extrusion are
(a) Magnesium (b) Aluminium
(c) Lead (d) All of these

47. The extrusion in which direction of motion of ram and extruded product is opposite
(a) Forward extrusion (b) Backward extrusion
(c) Direct extrusion (d) Hydrostatic

48. In backward extrusion friction force acts
(a) Between billet and container
(b) At die container interface
(c) Absent everywhere
(d) None of these

49. Backward extrusion is not used generally because
(a) Low friction between billet and container
(b) Container is subjected to high radial stress
(c) Handling of extruded product is difficult
(d) All of these

50. The extrusion in which high pressure fluid is used to apply extrusion force
(a) Impact extrusion (b) Hydrostatic extrusion
(c) Direct extrusion (d) Indirect extrusion

51. In extrusion process mandrel is used
(a) To form the base of tube (b) To bend the tube
(c) To break the tube (d) To join the tubes.

52. Cold working process is
(a) Extrusion (b) Rolling
(c) Wire drawing (d) Forging

53. Wire drawing speed for very thin wire is
(a) 10 m/min (b) 100 m/min
(c) 50 m/min (d) 1800 m/min

54. Which process is required to restore the ductility of material
(a) Normalizing (b) Annealing
(c) Hardening (d) Quenching

55. Tube drawing without mandrel is called
(a) Tube jamming (b) Blanking
(c) Tube sinking (d) Punching

56. Sheet metal working process is
(a) Press working (b) Forging
(c) Extrusion (d) Rolling

57. Which component of press mounts the tool and accessories.
(a) Ram (b) Bed
(c) Cushion (d) Bolster plate

58. The process in which parts are produced without formation of chip
(a) Drilling (b) Press working
(c) Grinding (d) Milling

59. The operation in which punch out is scrap and remaining strip is the required product
(a) Punching (b) Blanking
(c) Lancing (d) Slitting

60. Trimming is
(a) Formation of holes
(b) Production of narrow strips
(c) Removal of unwanted material from parting plane
(d) None of these

61. The operation in which large numbers of closely spaced holes are produced
(a) Slitting (b) Perforating
(c) Shaving (d) Lancing

62. Which operation is used to remove the metal in small increments
(a) Lancing (b) Perforating
(c) Nibbling (d) Punching **[UPTU Dec. 2010]**

63. Female component of die and punch assembly is
(a) Punch (b) Die
(c) Pilots (d) Bolster plate

64. Function of bolster plate is to
(a) Remove the scrap (b) Hold the punch
(c) Reduce pressure (d) Locate and support die assembly

65. Which component is used to hold the punch and die in proper alignment.

(a) Punch (b) Die

(c) Guide post and bushing (d) Stripper plate

66. Use of stripper plate is

(a) Remove the scrap (b) Hold the punch

(c) To reduce pressure (d) Locate and support die assembly

D. Match the Following

1.

(A)	(B)
(i) Wavy edges	(a) Nonmetallic inclusion
(ii) Alligatoring	(b) Very small reduction per pass
(iii) Folds	(c) Roll bending
(iv) Laminations	(d) Nonuniform deformation

2.

(A)	(B)
(i) Swaging	(a) above recrystallization temperature
(ii) Hot working	(b) below recrystallization temperature
(iii) Cold working	(c) forging operation
(iv) Blanking	(d) Sheet metal operation

3.

(A)	(B)
(i) Perforating	(a) Removal of unwanted material
(ii) Slitting	(b) Production of narrow strips
(iii) Notching	(c) Closely spaced holes are produced
(iv) Trimming	(d) Removal of specified portion of metal towards the edge

4.

(A)	(B)
(i) Direct extrusion	(a) Alligatoring
(ii) Indirect extrusion	(b) Die shift
(iii) Rolling defect	(c) Direction of movement of ram and extruded product is opposite
(iv) Forging defect	(d) Direction of movement of ram and extruded product are same

5.

(A)	(B)
(i) Closed die forging	(a) Bolts
(ii) Upset forging	(b) Collapsible tubes
(iii) Impact extrusion	(c) Leaf spring
(iv) Roll forging	(d) Flashless forging

6.

(A)	(B)
(i) Wire drawing	(a) Shear force
(ii) Extrusion	(b) Tensile force
(iii) Blanking	(c) Compressive force
(iv) Bending	(d) Spring back force

7.

(A)	(B)
(i) Drop forging	(a) Metal is gripped in the die and pressure is applied on heated end
(ii) Press forging	(b) Squeezing action

(iii) Upset forging	(c) Metal is placed between rollers and pushed
(iv) Roll forging	(d) Repeated hammer blow

8.

(A) Process	(B) Production of parts
(i) Rolling	(a) Discrete parts
(ii) Forging	(b) Rod and wire
(iii) Extrusion	(c) Solid and hollow parts
(iv) Drawing	(d) Flat plates and sheets

ANSWERS

A. True or False

1. False	2. True	3. False	4. True	5. False
6. False	7. True	8. False	9. True	10. False
11. True				

B. Fill in the Blanks

1. Recrystallization	6. Flakes
2. Hot	7. Tube drawing
3. Back up rolls	8. Press work
4. Laminations	9. Die
5. Closed die forging	10. Bolster Plate

C. Multiple Choice Questions

1. (d)	2. (c)	3. (a)	4. (c)	5. (a)
6. (c)	7. (b)	8. (c)	9. (b)	10. (c)
11. (d)	12. (b)	13. (c)	14. (a)	15. (b)
16. (b)	17. (b)	18. (a)	19. (d)	20. (d)
21. (b)	22. (b)	23. (d)	24. (b)	25. (c)
26. (b)	27. (b)	28. (d)	29. (a)	30. (a)
31. (b)	32. (c)	33. (c)	34. (d)	35. (b)
36. (c)	37. (d)	38. (c)	39. (a)	40. (b)
41. (c)	42. (c)	43. (a)	44. (b)	45. (a)
46. (d)	47. (b)	48. (b)	49. (c)	50. (b)
51. (a)	52. (c)	53. (d)	54. (b)	55. (c)
56. (a)	57. (d)	58. (b)	59. (a)	60. (c)
61. (b)	62. (c)	63. (b)	64. (d)	65. (c)
66. (a)				

D. Match the Following

1. (i)—(c), (ii)—(d), (iii)—(b), (iv)—(a)
2. (i)—(c), (ii)—(a), (iii)—(b), (iv)—(d)
3. (i)—(c), (ii)—(b), (iii)—(d), (iv)—(a)
4. (i)—(d), (ii)—(c), (iii)—(a), (iv)—(b)
5. (i)—(d), (ii)—(a), (iii)—(b), (iv)—(c)
6. (i)—(b), (ii)—(c), (iii)—(a), (iv)—(d)
7. (i)—(d), (ii)—(b), (iii)—(a), (iv)—(c)
8. (i)—(d), (ii)—(a), (iii)—(c), (iv)—(b)

❑❑❑

CHAPTER 6

Casting

INSIDE THIS CHAPTER

6.1. INTRODUCTION

It is one of the oldest manufacturing process and is first step in manufacturing of most of the products. In casting material is heated upto the desired temperature and converted into liquid form. The liquid metal is poured into mould cavity where it is solidified. After solidification product is taken out of mould cavity, cleaned and trimmed to get final shape.

6.2. CASTING TERMINOLOGY

(1) Pattern

It is replica of product to be cast in such a way that it can be used for forming an impression or cavity in moulding sand.

(2) Flask

It is used for holding the sand. It has two parts, upper moulding flask i.e. cope and lower moulding flask i.e. drag. Parting line divides the moulding flask in cope and drag.

(3) Core **[UPTU Dec. 2008]**

A core is a sand shape that is inserted into the mould to produce the cavity in a casting. Cores are shaped in core box and then baked or hardened.

(4) Core Print [UPTU Dec. 2008]

Core prints are extra projections provided on the pattern to form seats for cores in the mould and are used to support the core within the mould.

(5) Pouring Basin

It is a reservoir at the top of sprue through which molten metal is fed smoothly in sprue. It maintains the required rate of metal flow and reduces turbulence at sprue entry.

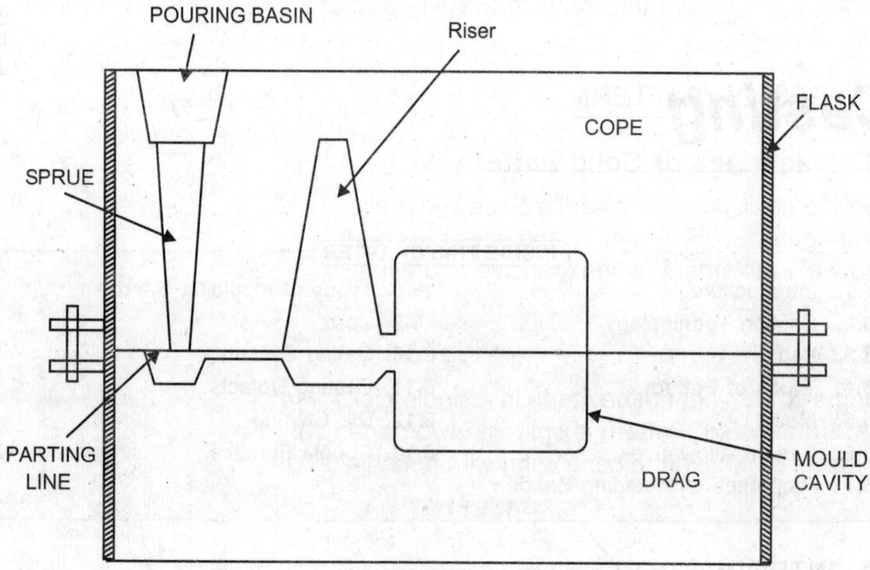

Fig. 6.1. Cross-section of Sand Mould.

(6) Riser [UPTU Dec. 2008]

It is a reservoir of molten material through which material flows into the mould cavity to compensate any shrinkage during solidification.

(7) Gate

Gate is the passage which delivers molten metal from runner to mould cavity. Size of gate depends upon solidification rate of molten metal.

(8) Runner [UPTU Dec. 2008]

It is a passageway through which molten metal is carried from sprue base to gates. It is generally located in drag.

(9) Draft

It is the taper on the pattern, which allows the easy removal of pattern from the mould.

(10) Sprue

Sprue is a tapered channel whose bigger end is connected to pouring basin and lower end to runner.

(11) Chills

Chills increase the cooling rate of casting and provide uniform or desired cooling rate.

6.3. PATTERN [UPTU DEC. 2008, MAY 2009]

Pattern is a replica of a product to be casted in such a way that it can be used for forming an impression or cavity in moulding sand. Pattern differs from the casting in the way that it carries allowances, draft and core prints. Materials used for making pattern are wood, metals, wax, plaster of paris, plastic, etc.

6.4. TYPES OF PATTERN

6.4.1. One Piece or Solid Pattern

Solid pattern is made in single piece and is used when shape is relatively simple and does not create withdrawal problems. It is inexpensive and usually made from wood.

Fig. 6.2. Solid Pattern.

6.4.2. Split Pattern

Some pattern cannot be made in a single piece due to difficulty in removing pattern from mould. Pattern is split into two parts by parting lines, in cope and drag. Tapered pins in the cope align with holes in drag for precise positioning.

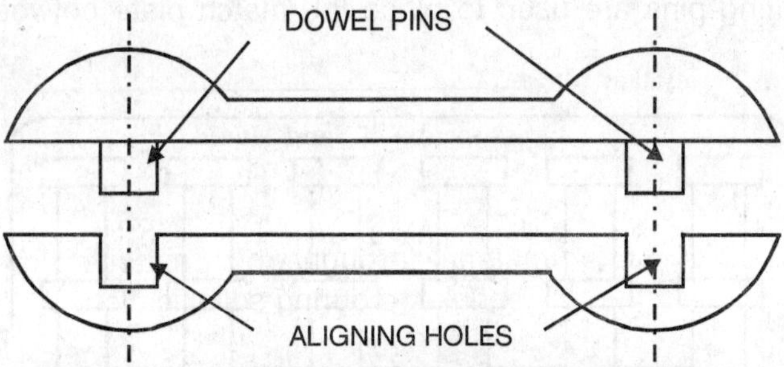

Fig. 6.3. Split Pattern.

6.4.3. Loose Piece Pattern

It is a pattern with loose piece to facilitate easy withdrawal of pattern from mould. First of all pattern is removed and after that loose piece are removed to form the required cavity. It is used to produce under cuts.

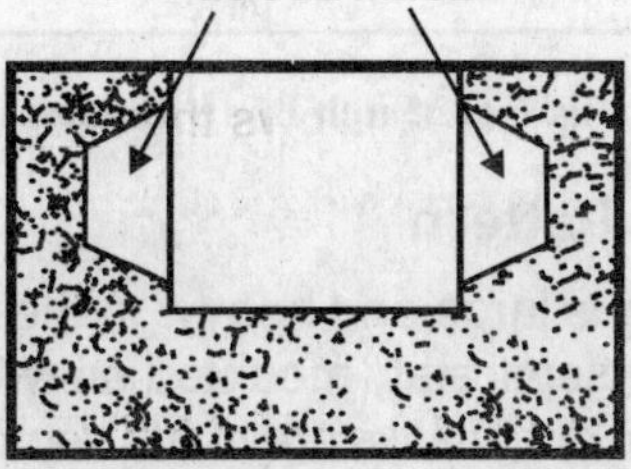

Fig. 6.4. Loose Piece Pattern.

6.4.4. Gated Pattern

Gated pattern consists of one or more loose pattern with attached gates and runners. These patterns eliminates the hand cutting of runner and gates thus resulting in improved productivity. Gated patterns are used for mass production of small castings.

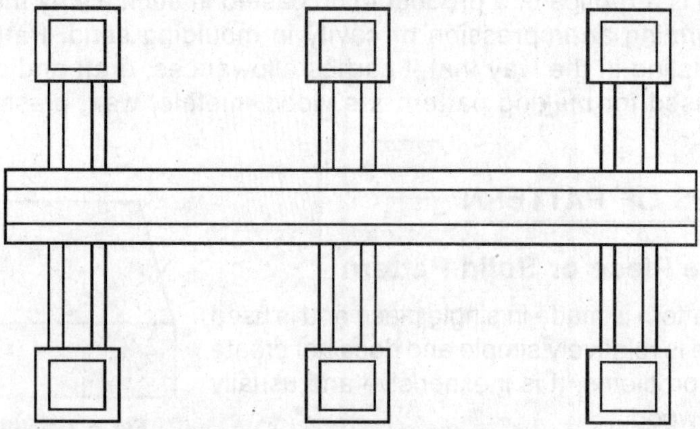

Fig. 6.5. Gated Pattern.

6.4.5. Match Plate Pattern

Cope and drag halves of a split pattern are permanently fastened to opposite sides of two match plates. It is used to produce large quantities of duplicate casting. Locating pins are used to place the match plate between cope and drag.

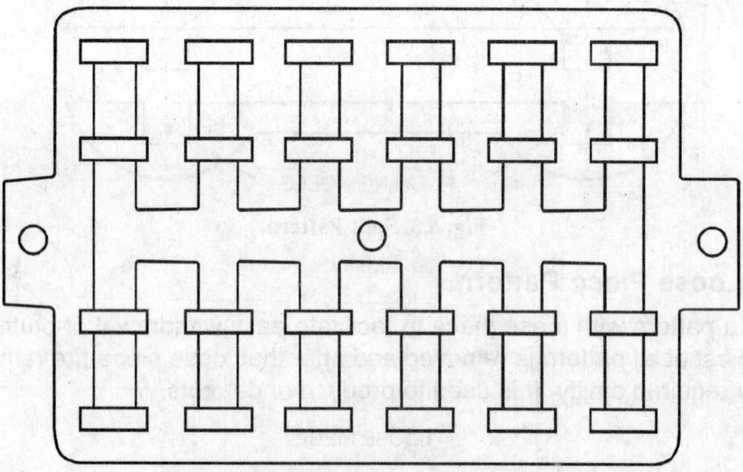

Fig. 6.6. Match Plate Pattern.

6.4.6. Cope and Drag Pattern

It is used when casting is large and heavy. In these patterns cope and drag halves of split patterns are separately mounted on two match plates to produce the couple mould. Cope and drag flasks are made separately and then assembled together to form complete cavity.

6.4.7. Sweep Pattern

Moulds which are symmetrical and have regular shape may be constructed by use of sweep pattern. Sweep pattern rotates about the post to get the desired shape of mould. This reduces the cost of three-dimensional patterns and is generally used for large castings.

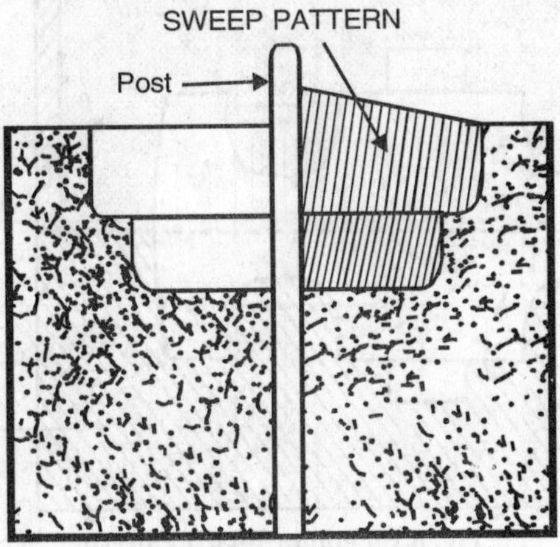

Fig. 6.7. Sweep Pattern.

6.4.8. Skeleton Pattern

Skeleton pattern is made up of strips of wood for outlining the shape of casting. These patterns are used for large castings having simple geometrical shape. It guides the moulder for hand shaping the mould.

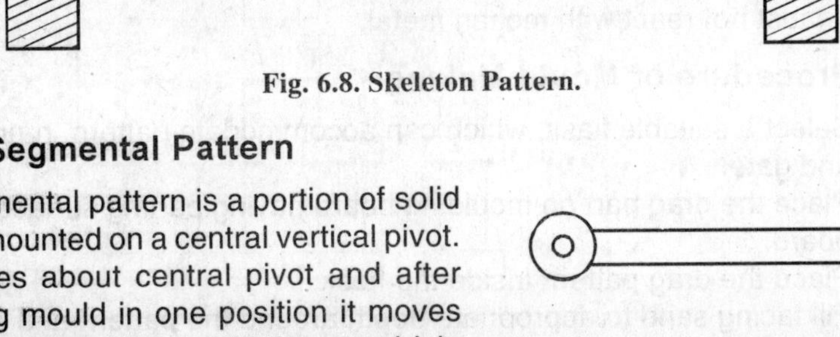

Fig. 6.8. Skeleton Pattern.

6.4.9. Segmental Pattern

Segmental pattern is a portion of solid pattern mounted on a central vertical pivot. It revolves about central pivot and after preparing mould in one position it moves to another position until the mould is complete. Such patterns are used to generate moulds of large circular castings such as rings, wheel rims and gears.

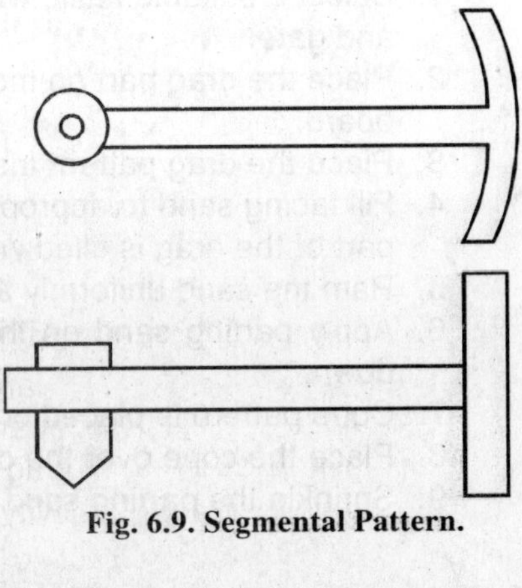

Fig. 6.9. Segmental Pattern.

6.4.10. Follow Board Pattern

A follow board is a wooden board which closely confirms the contour of fragile and structurally weak pattern to support it and avoid breaking or collapsing under the force of ramming.

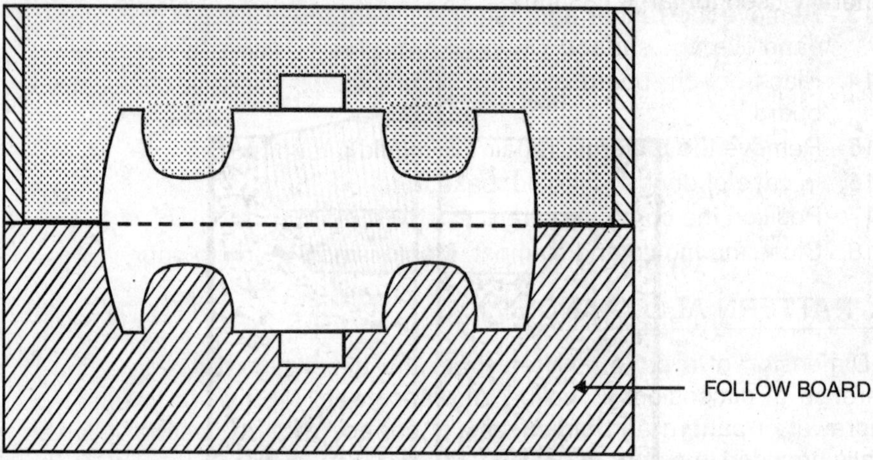

Fig. 6.10. Follow Board Pattern.

6.5. MOULD

A mould is a cavity produced in a compact sand mass due to withdrawal of pattern which when filled with molten metal produces a casting. The process of producing the cavity is known as moulding.

Characteristics of mould are :

1. A mould must possess enough refractoriness to withstand high temperature of molten metal.
2. A mould must be able to resist the erosive action of molten metal during pouring.
3. Mould must generate a minimum amount of mould gases to reduce defects.
4. It must possess sufficient strength to hold the weight of molten metal.
5. It must not react with molten metal.

6.5.1. Procedure of Mould Making

1. Select a suitable flask, which can accommodate pattern, runner, risers and gate.
2. Place the drag part on moulding board having parting surface down on board.
3. Place the drag pattern inside the flask.
4. Fill facing sand to appropriate depth around the pattern and rest of the part of the drag is filled with ordinary sand.
5. Ram the sand uniformly and remove the excess sand.
6. Apply parting sand on the top surface and turn the drag with upside down.
7. Cope pattern is placed on drag pattern in proper alignment with it.
8. Place the cope over the drag.
9. Sprinkle the parting sand around the cope pattern and over the drag.

10. Put runner and riser in appropriate position.
11. Fill facing sand to appropriate depth around the pattern and then fill cope with ordinary sand.
12. Ram the sand uniformly and remove the excess sand.
13. Remove the runner and riser pins, vent the cope and sprinkle parting sand over the surface.
14. Place bottom board over the cope and roll over the cope on bottom board.
15. Remove the patterns, repair the mould if required.
16. In case of dry sand mould, bake it.
17. Position the core if required.
18. Close the mould and clamp it. Mould is now ready to pour.

6.6. PATTERN ALLOWANCES [UPTU DEC. 2008]

Dimension of a pattern differs from the dimension of final casting. This difference of dimensions is called allowance and is provided to ensure easy withdrawal of pattern and dimensionally correct finished casting. Allowances usually provided in a pattern are :

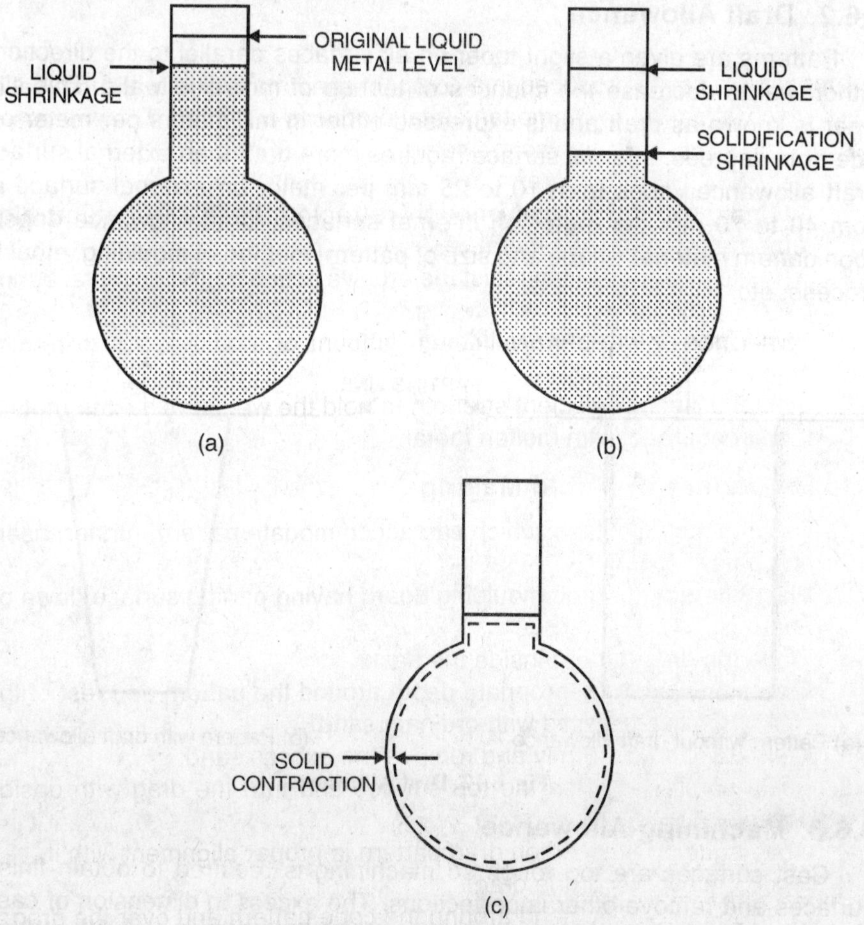

Fig. 6.11. Representation of Shrinkage.

6.6.1. Shrinkage Allowance

When hot metal cools from pouring temperature to room temperature it contracts. This contraction of cast metal takes place in three different steps (Fig. 6.11) :

(i) Contraction of liquid from pouring temperature to freezing temperature (liquid contraction).

(ii) Contraction due to change of phase from liquid to solid (solidifying contraction).

(iii) Contraction of casting from freezing temperature to room temperature (solid contraction).

First two contraction are compensated by risers and gates while contraction occurring in last stage is considered for shrinkage allowance. If the actual size of final casting is taken as dimension for pattern then the resulting casting will be smaller than desired due to contraction. So pattern is made of larger size as compared to required size of casting. For example, if a gear blank of 152 mm is to be casted then a pattern of 154 mm outside diameter should be taken to compensate for shrinkage.

6.6.2. Draft Allowance

Patterns are given a slight taper on all surfaces parallel to the direction of withdrawal to decrease the chances of tearing of moulding walls. This slight taper is known as draft and is expressed either in millimeters per meter on a side or in degrees. Internal surface requires more draft than external surfaces. Draft allowance varies from 10 to 25 mm per meter on external surface and from 40 to 70 mm per meter on internal surfaces. Draft allowance depends upon pattern material, shape and size of pattern, method of moulding, moulding process, etc.

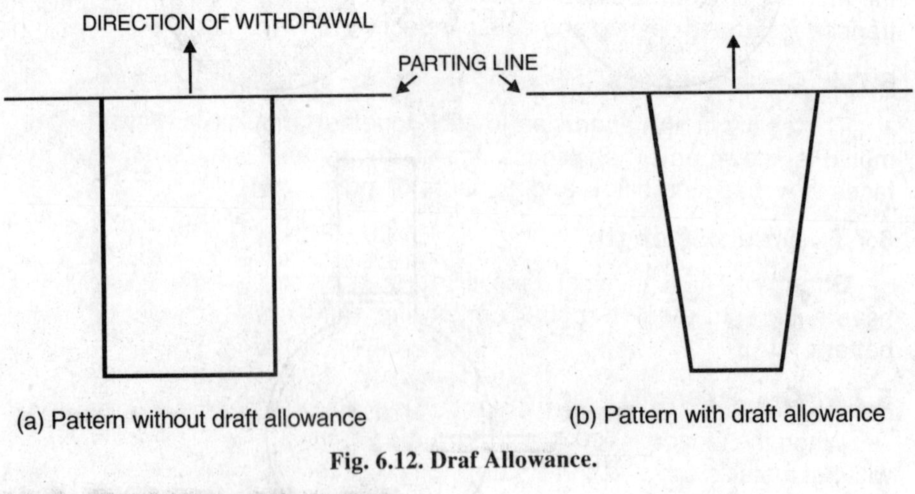

Fig. 6.12. Draf Allowance.

6.6.3. Machining Allowance

Cast surfaces are toọ rough so machining is required to obtain finished surfaces and remove other imperfections. The excess in dimension of casting other than that of final size to compensate for the loss of metal due to machining is known as machining allowance. Machining allowance depends on material, shape and size of casting, degree of surface finish, moulding process method of machining and functional requirement of casting.

6.6.4. Shake Allowance

When a pattern is rapped in the mould before its withdrawal, cavity in mould slightly increases and results in a enlarged final casting. So, it is desirable that original pattern dimension should be reduced to account for this increase. Shake allowance is provided only to large castings and considered as negligible in small and medium size castings.

6.6.5. Distortion Allowance

This is applicable to castings of irregular shape such as V and U sections. When such castings are allowed to cool they are distorted due to non-uniform contraction. To eliminate this, the shape of pattern should be given allowance of equal amount in opposite direction.

6.7. PROPERTIES OF MOULDING SANDS

There are various properties of sand that decide success of casting process.

6.7.1. Refractoriness

It measures the ability of moulding sand to withstand high temperature of molten metal without fusing. It increases with the grain size of sand.

6.7.2. Flowability

It is ability of sand to flow over and around the pattern and all portions of moulding flask during ramming. It increases by adding clay and water to sand.

6.7.3. Permeability

The ability of sand to allow gases to escape through the sand mould is known as permeability. During the solidification of casting gases escape from the mould and if sand is not having sufficient permeability these gases will be trapped inside the casting and cause defect.

6.7.4. Cohesiveness

The ability of sand particles to stick together and retain a given shape of mould is known as cohesiveness. It also imparts sufficient strength to mould faces to withstand the flow and pressure of molten metal.

6.7.5. Green Strength

Strength of sand in moist state is known as green strength. A mould should have sufficient green strength to retain the shape even after the removal of pattern.

6.7.6. Dry Strength

When the molten metal is poured in mould, the moisture of sand in contact with hot metal is converted in steam and makes the sand dry. So the strength of dry sand which enables it to resist erosion and withstand the pressure of molten metal is known as dry strength.

6.7.7. Hot Strength

Strength of mould in hot conditions generally above 1500°C is known as hot strength. Hot strength maintains the shape of mould cavity at high temperature and pressure of molten metal.

6.7.8. Adhesiveness

It is the property of moulding sand due to which sand particles are capable of adhesing to another body. Due to adhesiveness, the moulding sand mass clings to the sides of moulding base and does not fall out when the flask are lifted and turn over.

6.7.9. Collapsibility

It is the ability of sand to automatically collapse the sand mould after solidification of casting to permit free contraction of metal and thus preventing cracks and tears in casting. This property is very important for cores.

6.7.10. Chemical Resistivity

Due to this property, sand does not react chemically with molten metal and can be used repeatedly. Chemical resistivity also leads to good quality of cast surfaces.

Constituents of Moulding Sand

- Silica (SiO_2) - 86% to 90%
- Alumina (Al_2O_3) - 4% to 8%
- Iron Oxide (Fe_2O_3) - 2% to 5%
- Clay as a binder
- Moisture to provide elasticity

6.8. TYPES OF MOULDING SAND

6.8.1. Green Sand

It is mixture of silica sand with 18% to 30% clay and 6% to 8% water. It is generally used for rapid production of small or medium sized mould. Green sand mould requires less floor space as no storage is required. It is fine, soft, light, and porous. It can easily retain the shape given to it.

6.8.2. Dry Sand

Dry sand is a moulding sand from which excess moisture is removed by drying or baking. It is used for large castings. Dry sand mould provides the better surface finish and dimensional accuracy. Less venting is required as compared to green sand mould. It is more expensive than green sand.

6.8.3. Facing Sand

It possesses high strength and refractoriness. It consists of silica sand and clay without addition of used sand. It is used next to surface of pattern, which comes in contact with molten metal.

6.8.4. Loam Sand

It contains clay (up to 50%) and sand mixed with water. It has adhesive property so to hold on vertical surfaces of mould. Before casting it is dried slowly. Large regular shape such as chemical pans, drums are made by loam sand.

6.8.5. Backing Sand

It is used sand and forms back of facing layer. Due to burning in contact with molten metal and addition of coal dust its colour is black.

6.8.6. Parting Sand

Parting sand is sprinkled on the pattern and parting surfaces of the mould to avoid sticking of sand of one flask to other flask or to the pattern. Dry sand and burnt sand are used as parting sand.

6.8.7. Core Sand

Core sand is used for making cores and has high silica content. In this sand silica is mixed with linseed oil, resin, light mineral oil and other binders to impart high strength.

6.8.8. System Sand

System sand is used to fill the whole flask in machine moulding. It possess higher strength, permeability and refractoriness than backing sand.

6.9. CORE [UPTU MAY 2009]

Core is a sand shape generally perpared separately in a core box and baked in an oven. It is used to form internal cavities and hollow projections that are not possible to produce by pattern.

Cores are surrounded by hot molten metal and are subjected to more severe thermal stresses than the mould. So, the core should have sufficient strength and hardness to withstand high stresses and to support its own weight.

To make the core strong core binders are used with silica sand. Common binders used in core sand are molasses, resins, dextrin, core oil, linseed oil, etc. Organic binders develop strength by means of polymerization and cross linking. Core sand must also possess good refractoriness, permeability and collapsibility.

6.9.1. Types of Core

Cores are classified as

(i) On the Basis of Type of Sand Used

(a) **Green Sand Core :** These cores are formed by the pattern itself during moulding and are made with same sand used in making the mould. It has low strength and is used only for light castings.

(b) **Dry Sand Core :** These cores are prepared separately in core boxes, baked in an oven and then inserted in mould cavity. These core are placed after removal of pattern and before pouring of molten metal. Core sand is used to prepare these cores.

(ii) On the Basis of their Position

(a) **Horizontal Core :** When core is assembled in the mould by keeping its axis horizontal is called as horizontal core. It is usually cylindrical in shape and is positioned along the parting line with its one half in the cope and other half in drag. It is used for making through holes.

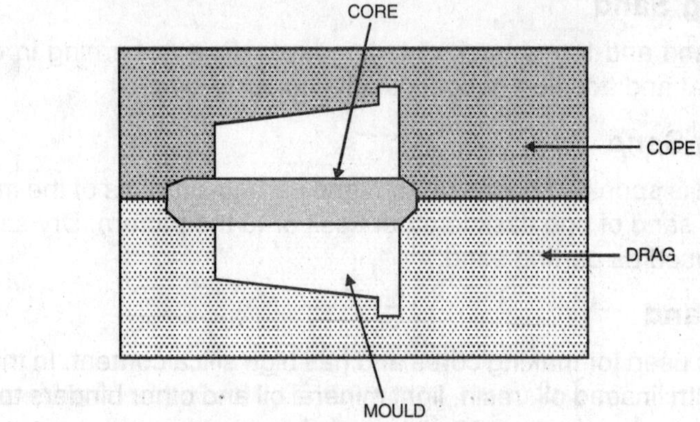

Fig. 6.13. Horizontal Core.

(b) **Vertical Core :** It is similar in shape to that of horizontal core but it is assembled in mould with its axis vertical.The ends of the core are placed in cope and drag.

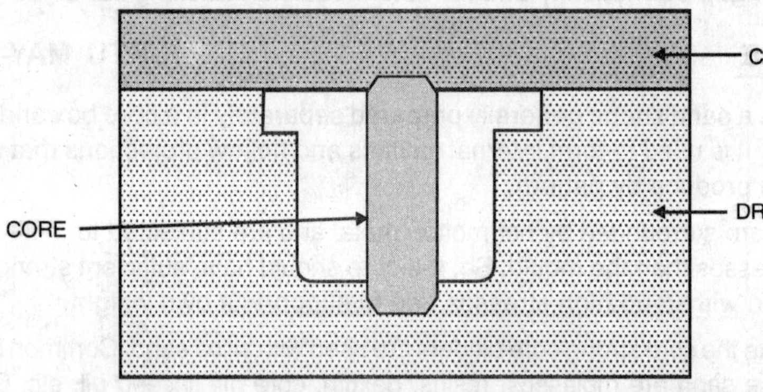

Fig. 6.14. Vertical Core.

(c) **Balanced Core :** Balanced core is a horizontal core supported only at one end and is used to produce a blind hole along a horizontal axis in casting.

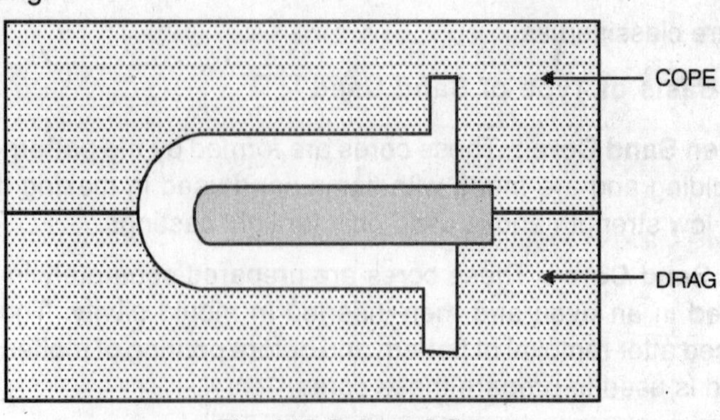

Fig. 6.15. Balanced Core.

(d) **Hanging Core :** Hanging core are those core which hangs at its top end and does not have any support at bottom in the drag.

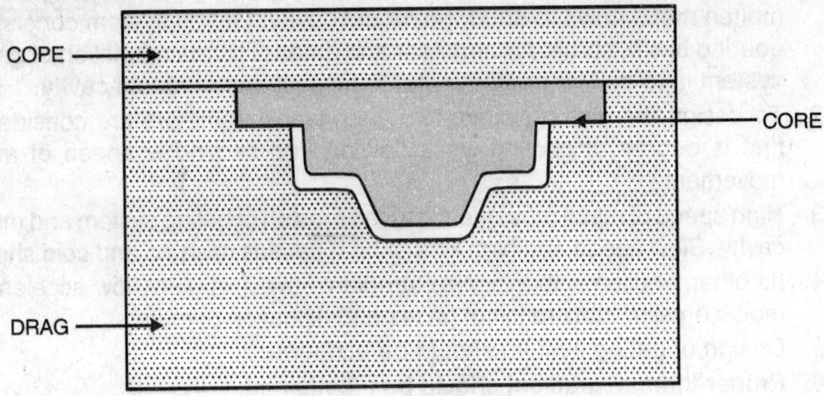

Fig. 6.16. Hanging Core.

(e) **Wire Core :** Wire core are used to produce hole in the casting either above or below the parting line. One side of wire core remain in level with inner surface of the mould so it is also known as stop-off core.

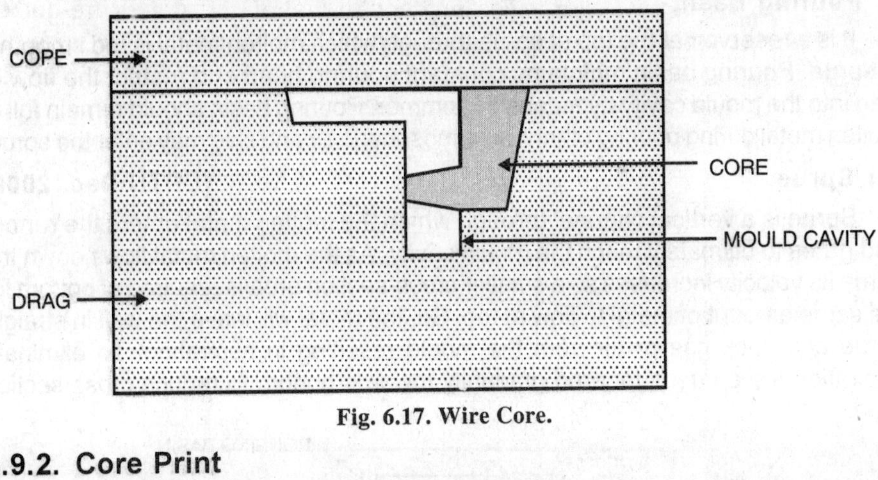

Fig. 6.17. Wire Core.

6.9.2. Core Print

Core print are extra projections provided on the pattern to position the core at proper place in mould cavity. Core print form the core seats to support cores such as horizontal, vertical, etc.

6.9.3. Difference Between Core and Core Print

Core	Core Print
1. Core is used to form cavity in a casting.	1. Core print are used to produce core seat to support the core.
2. Core is made in core box by core sand.	2. Core print are made on pattern with pattern material.
3. One core can produce only one casting.	3. One core print can be used in number of castings.
4. Core is a part of mould.	4. Core print is a part of pattern.

6.10. GATING SYSTEM

1. Gating system refers to all the channels and passageways through which molten metal flows to enter the mould cavity. Gating system consists of pouring basin, sprue, runner, riser and gates. The main function of gating system is to deliver molten metal to all sections of mould cavity.
2. To design the gating system two important parameters are considered, first is degree of cooling while flowing and second is speed of metal movement.
3. High speed of metal movement causes erosion of gating system and mould cavity. Slow speed and high loss of heat causes misruns and cold shots.
4. Its other function is to minimize turbulent flow. Turbulent flow accelerates mould erosion, absorption of gases and oxidation of metal.
5. Design of gating system should be economical.
6. Proper thermal gradient should be maintained.
7. Slag, dross and other unwanted material should not be allowed.
8. Casting yield should be maximized.

6.10.1. Elements of Gating System

(i) Pouring Basin

It is a reservoir at the top of sprue through which molten metal is fed smoothly in sprue. Pouring basin maintains constant pouring head and restrict the flow of slag into the mould cavity by means of skimmer. Pouring basin should remain full of molten metal during pouring otherwise atmospheric air and slag may enter the sprue.

(ii) Sprue **[UPTU Dec. 2008]**

Sprue is a vertical channel through which the molten metal enters the runner and gates to ultimately reach the mould cavity. As the molten metal flows down the sprue its velocity increases so a smaller cross-section area is required at bottom for the same amount of metal to flow at top. So, the metal will leave the wall in straight sprue and suck gas or air from the mould resulting in aspiration. To eliminate aspiration the sprue is tapered gradually towards bottom to reduce cross-section area.

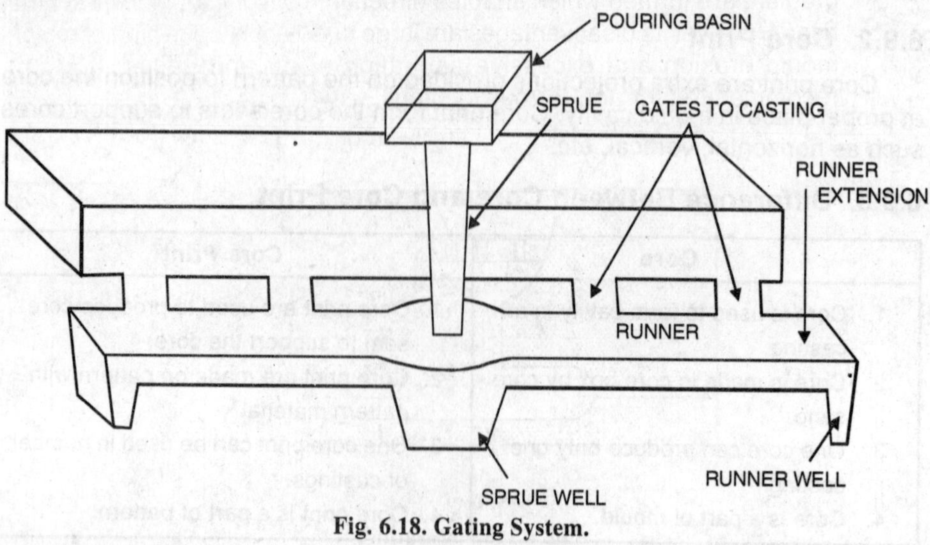

Fig. 6.18. Gating System.

(iii) Sprue Base Well

Sprue base well is a reservoir provided at the bottom of sprue to reduce the momentum of falling molten metal and to absorb kinetic energy as flow changes its direction from vertical to horizontal. Sprue base well reduces the mould erosion by making the flow more uniform in runner.

(iv) Runner

It is a passage through which molten metal reaches from sprue base to gates around the cavity. Runners are generally provided in drag but depending upon the shape of casting it may be located in cope. Runner should flow full for effective trapping of slag.

(v) Runner Extension

Runner extended a little further beyond the ingate is known as runner extension. Metal alongwith slag flows straight and reaches beyond the ingate where the slag is trapped in extension. Runner extension also absorbs the kinetic energy and causes a smooth flow of metal in runner and mould cavity.

(vi) Gate

Gates are channels which connect the runners to the mould cavity and through which molten metal flows to fill a mould cavity. The size and location of gate depends upon number of factors such as

(a) Flow rate of metal.
(b) Distribution of metal in mould cavity.
(c) Filling of mould.
(d) Erosion of mould.
(e) Temperature loss.
(f) Removal of gate without damaging the casting.

Types of Gates

(a) Top Gate : In this type of gates, molten metal enters from the top in mould cavity. The advantage of top gating is that proper temperature gradient are formed which enables directional solidification towards the riser. However its disadvantages are large turbulence, splashing of metal, mould erosion and excessive aspiration of air and gases. It is not recommended for non-ferrous castings.

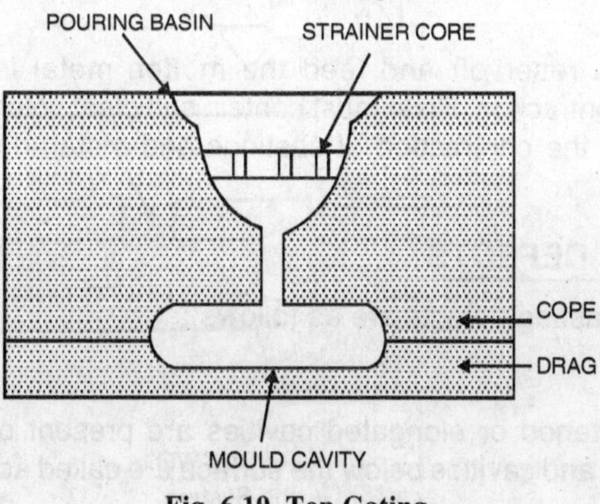

Fig. 6.19. Top Gating.

(b) Bottom Gate : In bottom gates, molten metal enters from the bottom of mould cavity and rises slowly and steadily in the mould. These gates maintains the turbulence of metal at minimum level and thus prevents mould erosion. Drawback of these gates are unfavourable temperature gradient and difficult to achieve directional solidification especially when riser is at the top of casting.

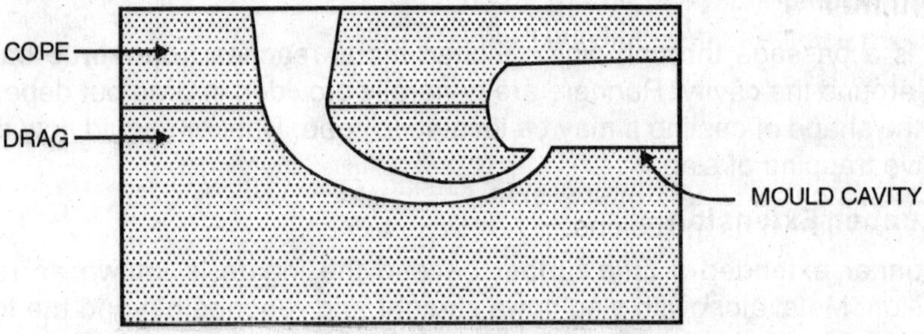

Fig. 6.20. Bottom Gating.

(c) Parting Gate : In parting line gate, molten metal enters the mould at the parting line. It is inexpensive and has characteristic of both systems, i.e. it is a top gate for the mould cavity in drag and a bottom gate for cavity in cope. These gates are most common and most satisfactory except for very deep moulds which have considerable distance from the gate to the bottom of mould cavity.

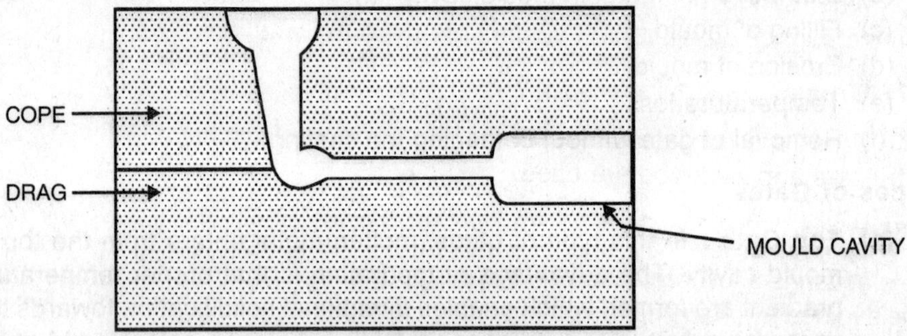

Fig. 6.21. Parting Gate.

(vii) Riser

Riser acts as reservoir and feed the molten metal in mould cavity to compensate for contraction. Riser must contain sufficient volume of molten metal for compensating the contraction of castings and metal in riser must solidify after the casting.

6.11. CASTING DEFECTS

The various casting defects are as follows :

(1) Blow Holes

Rounded, flattened or elongated cavities are present on the surface are called open blows and cavities below the surface are called as blow holes. When the molten metal is poured into cavity, moisture of the mould is converted into steam. Part of this steam is trapped in casting resulting in open blows or blow holes. Proper ventilation is required to avoid this defect.

(2) Pin Holes

Pin holes are caused due to presence of hydrogen in molten metal. When hydrogen escapes the solidifying metal pin holes are formed at or just below the casting surface.

(3) Inclusion

Undesirable foreign material such as slag, oxides, dirt etc. present in casting metal are known as inclusions.

(4) Gas Holes

This defect refers to entrapped gas bubbles of nearly spherical shape resulting due to large amount of gases dissolved in liquid metal.

(5) Porosity

Decrease in gas solubility during solidification results in porosity due to which very small holes are dispersed in the casting.

(6) Drop

When sand drops from the cope into the mould, an irregular projection occurs on the surface called drop. Gaggers are used to avoid this.

(7) Dross

Dross are light impurities at the top surface of casting. To avoid dross, strainers and skim bob are used.

(8) Wash

Wash is a low projection on the drag surface of casting near to gate. It can be avoided by changing the gating design to reduce the turbulence and selecting a proper moulding sand.

(9) Buckle

A long, fairly shallow, broad, V shaped depression in the surface of flat castings is known as buckle. It occurs due to expansion of sand by hot metal.

(10) Swell

Due to high liquid metal pressure the moulding wall is pushed outward at certain places resulting in an enlargement of mould cavity. This additional space is filled by molten metal producing the swell defect. To remove this defect proper ramming is required.

(11) Hot Tear

High residual stress causes crack in the casting called hot tears. Main reason for this defect is poor design of casting.

(12) Misrun

Misrun defect occurs when the molten metal is not able to fill the entire mould cavity thus leaving unfilled cavities. Too small casting thickness, too low pouring temperature and too low fluidity of molten metal causes this defect.

(13) Cold Shut

Cold shut occurs when two streams of metal coming from opposite direction do not fuse together due to cooling of molten metal. Too low fluidity of molten metal causes this defect.

(14) Run Out

Run out occurs due to leakage of molten metal from the mould cavity. Main causes of this defect are faulty moulding and defective moulding boxes.

(15) Scab

Scab are sort of projections which occurs when a portion of mould face lifts and molten metal flows underneath in a thin layer. Scab occurs due to high moisture content in sand, too low permeability, too fine sand and uneven mould ramming.

(16) Rat Tail

Rat tail is a long, shallow, angular depression resulting due to slight compression failure of the thin layer of moulding sand.

(17) Scar

Scar is a shallow blow which usually occurs on a flat surface of casting.

(18) Blister

Blister is a scar covered by thin layers of a metal.

(19) Fin

A thin unintended projection of metal usually occurring at the parting line is known as fin. It occurs due to improper clamping, over flexible bottom boards and inadequately weighted sand.

(20) Shift

Misalignment between two halves of a mould or of a core results in mould shift or core shift.

6.12. DIE CASTING

Die casting is a precision casting technique in which liquid metal is injected into metallic die under high pressure. It is used for production of fine surface, complex shapes and narrow sections. Die casting is done by using hot chamber die casting machine or cold chamber die casting machine.

6.12.1. Hot Chamber Die Casting Machine

In hot chamber die casting machine, furnace is integral with casting machine to melt the metal in a holding pot fitted in furnace. Goose neck plays an important role in supplying molten metal in the die cavity and is submerged in holding pot containing molten metal. At the top portion of gooseneck a port is provided to allow the molten metal inside the cylinder. During the downward motion of plunger the port gets closed and then the plunger applies the pressure on molten metal to flow through injector in die casting. After a certain time piston moves upward due to which unused molten metal returns into gooseneck. Just before the end of upward stroke port is uncovered letting the molten metal to flow into the gooseneck. This method is suitable for zinc, tin, lead and other low melting point alloys.

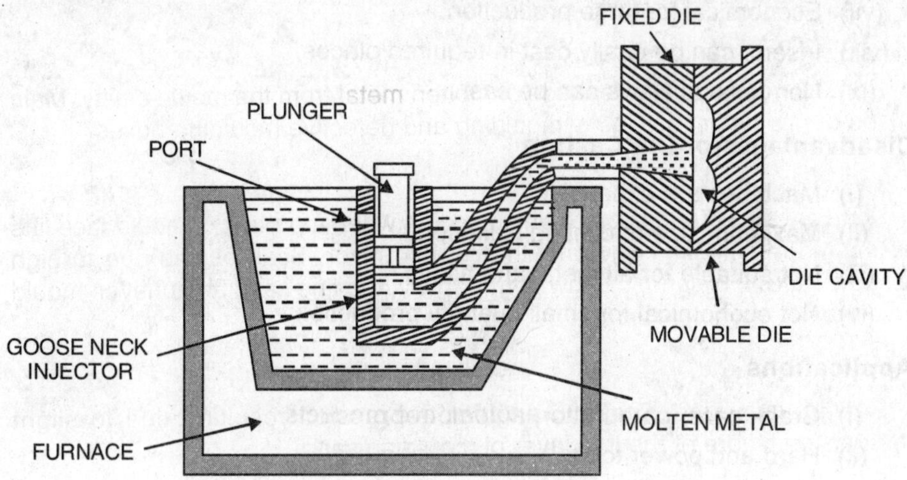

Fig. 6.22. Hot Chamber Die Casting Machine.

6.12.2. Cold Chamber Die Casting Machine

In hot chamber process it is difficult to cast materials having high melting temperature and to avoid this problem cold chamber process is used in which furnace is not integral with machine.

In this process ladle is used to fill the molten metal into the chamber for every shot. Plunger reciprocates in horizontal cylinder and apply pressure on molten metal to flow into die cavity to form the casting. It is suitable for high melting point alloys such as aluminium alloys and brass but its production rate is less as compared to hot chamber machine.

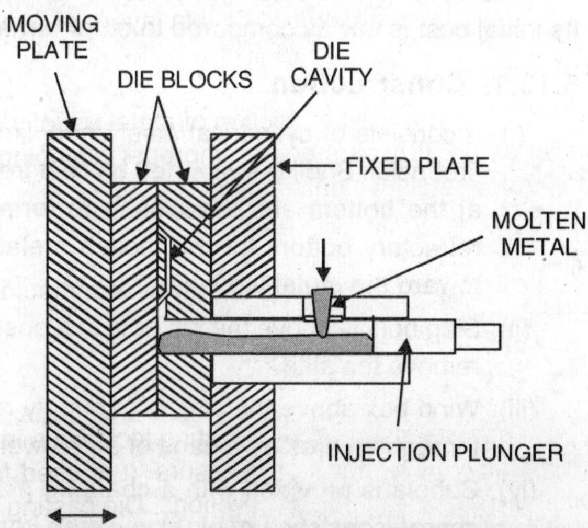

Fig. 6.23. Cold Chamber Die Casting Machine.

Advantages of Die Casting

(i) High production rate.
(ii) Close dimensional tolerance can be achieved.
(iii) Small holes can be cored.
(iv) Die has long life.
(v) Very good surface finish can be obtained.
(vi) Thin section can be cast.
(vii) Economical for large production.
(viii) Inserts can be easily cast in required places.
(ix) Non-ferrous alloys can be cast.

Disadvantages of Die Casting

(i) Machines are expensive.
(ii) Maximum size of casting is limited.
(iii) Not suitable for all metal and alloys.
(iv) Not economical for small quantity production.

Applications

(i) Crank case, carburettor, automotive products.
(ii) Hard and power tools.
(iii) Pump parts, plumbing parts, etc.

6.13. CUPOLA FURNACE

Cupola furnace is extensively used in its various forms for melting cast iron. Its initial cost is low as compared to other furnaces.

6.13.1. Construction

(i) It consists of cylindrical steel shell, lined internally with heat resisting firebricks. Shell is supported on cast iron legs. Drop doors are provided at the bottom and are closed to form the sand bed which act as a refractory bottom for the molten metal and coke. This bottom slopes toward the metal tap hole.
(ii) Slag hole is above the tap hole in a position opposite to it and is used to remove the slag.
(iii) Wind box above the slag hole supply required quantity of air to cupola through tuyeres by means of air blower.
(iv) Cupola is provided with a charging platform and charging door to fed charge consisting of pig iron, scrap iron, coke and fluxes.
(v) At the top of furnace, spark arrester is provided to arrest the spark and allow only smoke to escape from opening.

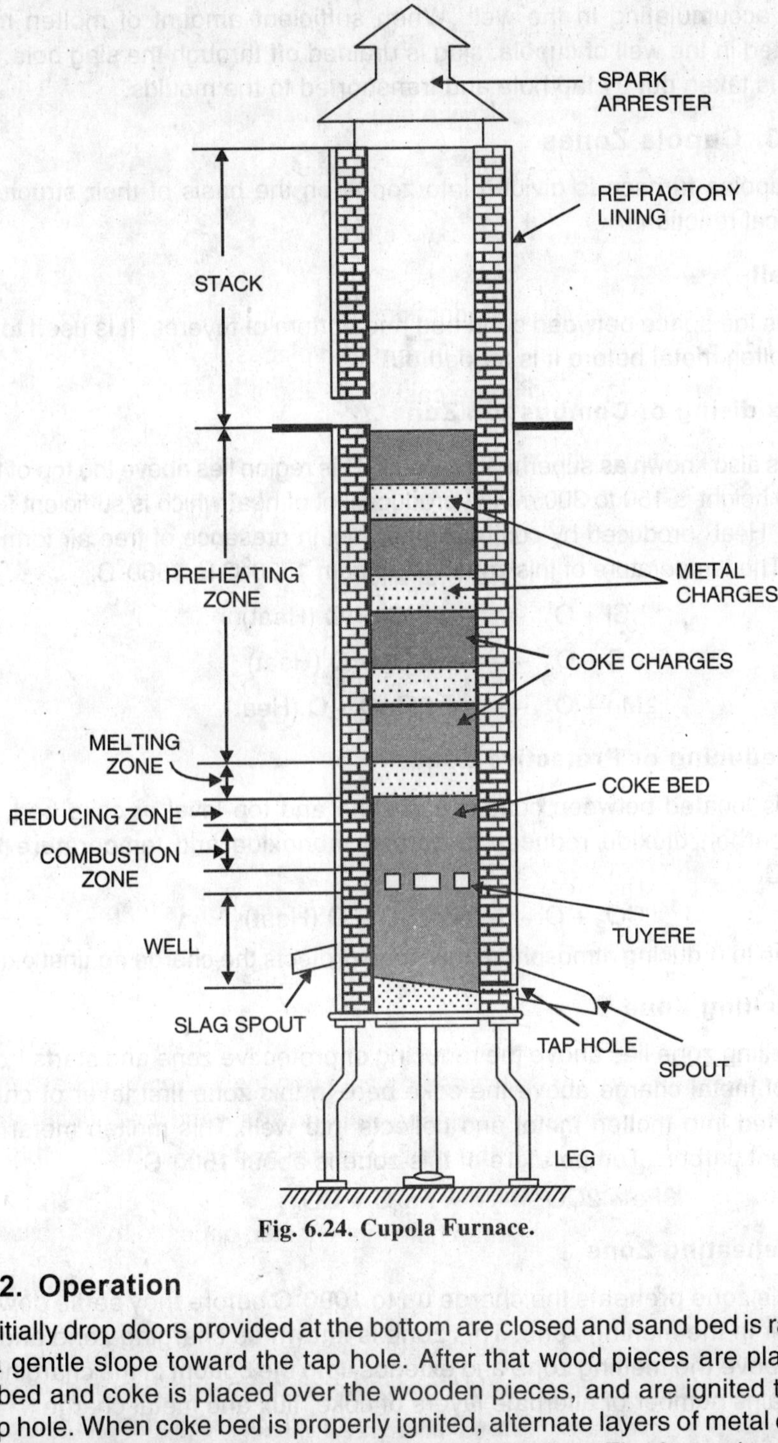

Fig. 6.24. Cupola Furnace.

6.13.2. Operation

Initially drop doors provided at the bottom are closed and sand bed is rammed with a gentle slope toward the tap hole. After that wood pieces are placed on sand bed and coke is placed over the wooden pieces, and are ignited through the tap hole. When coke bed is properly ignited, alternate layers of metal charge, flux and coke are fed into cupola through the charge door. Flux is used to remove the oxides and other impurities present in metal in form of slag.

The charge is then allowed to soak the heat and after some time full blast is turned on. After the blast has been opened for about 10 minutes, molten metal starts accumulating in the well. When sufficient amount of molten metal is collected in the well of cupola, slag is drained off through the slag hole. Molten metal is taken out by tap hole and transported to the moulds.

6.13.3. Cupola Zones

Cupolas furnace is divided into zones on the basis of their structure and chemical reactions :

(i) Well

It is the space between sand bed and bottom of tuyeres. It is used to collect the molten metal before it is tapped out.

(ii) Oxidizing or Combustion Zone

It is also known as superheating zone. This region lies above the top of tuyeres and its height is 150 to 300 mm. It produces lot of heat which is sufficient for other zones. Heat produced by combustion of fuel in presence of free air form the air blast. The temperature of this zone varies from 1530°C to 1860°C.

$$Si + O_2 \longrightarrow SiO_2 + Q \text{ (Heat)}$$

$$C + O_2 \longrightarrow CO_2 + Q \text{ (Heat)}$$

$$2Mn + O_2 \longrightarrow 2MnO + Q \text{ (Heat)}$$

(iii) Reducing or Protective Zone

It is located between combustion zone and top level of coke bed. In this zone carbon dioxide reduces to carbon monoxide and temperature falls to 1200°C.

$$CO_2 + C \longrightarrow 2CO - Q \text{ (Heat)}$$

Due to reducing atmosphere this zone protects the charge against oxidation.

(iv) Melting Zone

Melting zone lies above the reducing or protective zone and starts from first layer of metal charge above the coke bed. In this zone first layer of charge is converted into molten metal and collects into well. This molten metal attains sufficient carbon. Temperature of this zone is about 1600°C.

$$3Fe + 2CO \longrightarrow Fe_3C + CO_2$$

(v) Preheating Zone

This zone preheats the charge up to 1090°C before they settle downward to enter in the melting zone. This zone is known as charging zone and starts from above the melting zone and extends upto the bottom of the charging door. It contains number of alternate layers of coke, flux and metal charge.

(vi) Stack

It lies above the preheating zone and provides the passage for hot gases generated within the furnace.

SUMMARY

- In casting material is heated upto the desired temperature and converted into liquid form. The liquid metal is poured into mould cavity where it is solidified. After solidification product is taken out of mould cavity, which is cleaned and trimmed to get final shape.
- Pattern is replica of product to be casted in such a way that it can be used for forming an impression or cavity in moulding sand.
- Riser is a reservoir of molten material through which material flows into the mould cavity to compensate any shrinkage during solidification.
- Draft is the taper on the pattern, which allows the easy removal of pattern from the mould.
- Cores are used to make internal cavities and hollow projection that are not possible to produce by pattern.
- Core prints are extra projections provided on the pattern to form seats for core in the mould and are used to support the core within the mould.
- Runner is a passageway through which molten metal is carried from sprue base to gates.
- Gate is defined as opening through which molten metal enters into the mould cavity.
- Die casting is a precision casting technique in which liquid metal is injected into metallic die under high pressure. It is of two type, hot chamber die casting machine and cold chamber die casting machine.

SHORT QUESTIONS WITH ANSWERS

Q.1. What is the use of riser in casting? **[UPTU May 2012]**

Ans. Riser acts as reservoir which supply the molten metal to die cavity to compensate any shrinkage during solidification.

Q.2. Define pattern.

Ans. Pattern is a replica of the product to be cast in such a way that it can be used for forming an impression or cavity in moulding sand.

Q. 3. Why draft allowance is provided in casting?

Ans. Draft allowance is provided for easy withdrawal of pattern from mould and to prevent damage of edges and surfaces of mould.

Q. 4. What are the main constituents of green sand?

Ans. Green sand is mixture of silica sand, 18 to 30% clay and 6 to 8% water.

Q. 5. What is the use of core in casting?

Ans. Cores are used to make internal cavities and hollow projections.

Q. 6. What are two important parameters while designing the gating system?

Ans. (1) Degree of cooling

(2) Speed of metal movement.

Q. 7. How the blow holes can be avoided in casting?

Ans. By providing proper ventilation blow holes can be avoided.

Q. 8. Write any four defects of casting. **[UPTU May 2010]**

Ans. (1) Porosity

(2) Hot tear

(3) Buckle

(4) Run out

Q. 9. What are the applications of die casting?

Ans. Crank case, Magnets, Carburettor, Decorative items.

Q.10. Why the flux is used in Cupola furnace?

Ans. Flux is used to remove the oxide and other impurities present in metal in form of slag.

Q. 11. Write two properties of moulding sand. **[UPTU May 2010]**

Ans. (1) Refractoriness

(2) Green strength

Q. 12. What is the product of cupola furnace. **[UPTU May 2012]**

Ans. Cast iron

Q. 13. Define Gating ratio. **[UPTU Dec. 2011]**

Ans. Gating ratio is defined as ratio of sprue area to total runner area to total gate area.

REVIEW QUESTIONS

1. Explain different type of moulding sands.
2. Explain the properties of moulding sand. **[UPTU Dec. 2010]**
3. Explain the procedure of making mould.
4. Discuss various allowances provided in castings.
5. What is the function of core? Explain different types of core.
6. Explain construction and operation of cupola furnace and its limitations. **[UPTU Dec. 2009]**
7. Compare hot and cold chamber die casting machine.
8. Explain the working of hot and cold chamber die casting machine with neat sketches.
9. Explain different types of casting defects.
10. Define pattern? Explain different types of pattern.
11. Explain different components of gating system and classify different types of gates.
12. Explain : **[UPTU Dec. 2008, May 2011]**
    (a) Mould (b) Pattern
    Describe the two important pattern allowances.
13. Define the following terms with sketch as used in sand casting
    (a) core (b) core prints (c) sprue (d) runner (e) riser. **[UPTU Dec. 2008]**
14. Give the casting terminology with neat sketch.
15. Why the pattern is different from casting. Describe with neat sketch, the various steps in casting process. **[UPTU May 2010]**

SELF TEST QUESTIONS

A. State Whether True or False

1. In cold chamber die casting machine furnace is integral with casting machine.
2. Port provided on top of goose neck in die casting machine supplies fuel to chamber.

3. Core print is used to locate the pattern.
4. Facing sand is used sand and forms back of facing layer.
5. Dross are light impurities at top surface of casting.

B. Fill in The Blanks

1. Pattern is a .... of product to be cast.
2. Hole is produced in casting with the help of .... **[UPTU Dec. 2009]**
3. Chill .... the cooling rate of casting.
4. Allowance provided on surfaces parallel to direction of withdrawal is known as ....
5. Refractoriness is the ability of moulding sand to withstand high temperature without ....
6. Core prints are used to support the .... in mould cavity.
7. Blow holes are produced by..............
8. Hot tears are cracks caused due to ....

C. Multiple Choice Questions

1. Upper part of moulding flask is known as
   (a) Cope (b) Drag
   (c) Chaplet (d) Pattern
2. To provide shrinkage allowance pattern is made
   (a) Larger (b) Smaller
   (c) Actual size (d) None
3. Balanced core is supported at
   (a) Both end (b) One end
   (c) Throughout its length (d) None
4. Pouring basin is used to avoid
   (a) Oxidation of metal (b) Rapid cooling
   (c) Mould erosion (d) None
5. Good temperature gradient is achieved in
   (a) Top gate (b) Bottom gate
   (c) Parting gate (d) None
6. Mould erosion is less in
   (a) Top gate (b) Bottom gate
   (c) Parting gate (d) None
7. Inclusions are **[UPTU Dec. 2010]**
   (a) Iron particle (b) Steel particle
   (c) Non metallic particle (d) None
8. Drop is a
   (a) Light impurities (b) Moulding sand
   (c) Irregular projection (d) Cavity
9. Temperature of oxidizing or combustion zone varies from
   (a) 1530°C to 1860°C (b) 1200°C to 1400°C
   (c) 1000°C to 1160°C (d) 1830°C to 2060°C
10. Sweep pattern are used for
    (a) Irregular shape mould (b) Symmetrical shape mould.
    (c) Large mould (d) Small mould

11. Blow holes are common defects found in **[UPTU Dec. 2008]**

(a) Welding (b) Machining
(c) Forming (d) Casting

12. Which is used to produce core seat to support the core

(a) Core (b) Core print
(c) Chills (d) Riser

13. Replica of product is known as

(a) Mould (b) Flask
(c) Pattern (d) Core

14. The process in which product is obtained by pouring molten metal into mould cavity

(a) Forging (b) Casting
(c) Machining (d) Welding

15. _______ is used to produce the cavity in a casting.

(a) Pattern (b) Flask
(c) Cope (d) Core

16. ______ are extra projections provided on the pattern to form seats for core in the mould.

(a) Core (b) Core print
(c) Gate (d) Runner

17. Draft is provided on pattern for

(a) Easy removal of pattern from mould
(b) Easy removal of flask from mould
(c) Easy removal of runner from mould
(d) Easy removal of core from mould

18. ______ increase the cooling rate of casting and provide uniform cooling rate.

(a) Sprue (b) Chills
(c) Draft (d) Core

19. Which material is used for making pattern

(a) Wood (b) Wax
(c) Plastic (d) All of these

20. ______ pattern consist of one or more loose pattern with attached gates and runners.

(a) Match plate (b) Cope and drag
(c) Gated (d) Sweep

21. Pattern used to produce large quantities of duplicate casting is

(a) Loose piece pattern (b) Sweep pattern
(c) Split pattern (d) Match plate pattern

22. Pattern made of strips of woods for outlining the shape of casting is

(a) Cope and drag pattern (b) Sweep pattern
(c) Skeleton pattern (d) Match plate pattern

23. ______ pattern used to generate moulds of large circular castings such as rings, gears is

(a) Segmental (b) Follow board
(c) Skeleton (d) Match plate pattern

24. Sand cavity produced by withdrawal of pattern, is called
(a) Cope (b) Core
(c) Mould (d) Drag

25. Difference in the dimension of pattern and casting is called
(a) Clearance (b) Allowance
(c) Fit (d) Tolerance

26. To compensate for shrinkage, pattern is made of ______ size as compared to required size of casting
(a) Larger (b) Smaller
(c) Equal (d) None of these

27. Slight taper provided on pattern surface parallel to direction of withdrawal, is known as
(a) Riser (b) Draft
(c) Chill (d) Runner

28. Operation required for finishing of cast surface is
(a) Forging (b) Welding
(c) Machining (d) Milling

29. To compensate for machining allowance, dimension of pattern is ______ than final dimension.
(a) Larger (b) Smaller
(c) Equal (d) None of these

30. The ability of moulding sand to withstand high temperature of molten metal without fusing is
(a) Permeability (b) Refractoriness
(c) Cohesiveness (d) Collapsibility

31. The ability of sand to allow gases to escape through the sand mould is
(a) Permeability (b) Refractoriness
(c) Cohesiveness (d) Collapsibility

32. Strength of sand in moist state is ________
(a) Dry strength (b) Hot strength
(c) Green strength (d) None of these

33. The ability of sand particles to stick together and retain a given shape of mould is
(a) Flowability (b) Collapsibility
(c) Adhesiveness (d) Cohesiveness

34. Moulding sand contain ______ silica
(a) 20 to 40% (b) 86 to 90%
(c) 43 to 56% (d) 58 to 68%

35. ________ is mixture of silica sand with 18 to 30% clay and 6 to 8% water
(a) Facing sand (b) Green sand
(c) Dry sand (d) Backing sand

36. Which sand forms back of facing layer.
(a) Loam sand (b) Core sand
(c) Facing sand (d) Backing sand

37. Parting sand is used to
(a) Make cores
(b) To hold vertical surfaces of mould

(c) To avoid sticking of sand of one flask to other flask
(d) To form back of facing layer

38. ______ are used to make the core strong,
(a) Water (b) Hard sand
(c) Core binders (d) None of these

39. Which core is used for making through holes
(a) Horizontal core (b) Vertical core
(c) Balanced core (d) Hanging core

40. ______ is used to produce a blind hole along horizontal axis in casting
(a) Horizontal core (b) Vertical core
(c) Balanced core (d) Hanging core

41. High speed of molten metal cause
(a) Erosion of gating system and mould cavity
(b) Cooling of molten metal
(c) Damage of flask
(d) Misruns and cold shots

42. The reservoir at the top of sprue through which molten metal is fed in sprue is
(a) Runner (b) Pouring basin
(c) Gate (d) Runner extension

43. Sprue is tapered gradually towards bottom to
(a) Removal of slag (b) eliminate aspiration
(c) To measure flow rate (d) All of these

44. _____ reduce the mould erosion by making the flow more uniform in runner
(a) Pouring basin (b) Sprue
(c) Gate (d) Sprue base well

45. The passage through which molten metal reaches from sprue base to gates is known as
(a) Riser (b) Runner
(c) Sprue (d) Runner extension

46. Channel which connect the runner to mould cavity is
(a) Pouring basin (b) Gate
(c) Sprue (d) Riser

47. Which type of gate is not preferred for nonferrous castings
(a) Top gate (b) Bottom gate
(c) Parting gate (d) None of these

48. Which gate maintain the turbulence of metal of minimum level and prevent mould erosion
(a) Top gate (b) Parting gate
(c) Bottom gate (d) None of these

49. Advantage of top gate is
(a) Large turbulene (b) Splashing of metal
(c) Mould erosion (d) Enable directional solidification

50. Which part of casting is used to feed the molten metal in mould cavity to compensate for contraction.
    (a) Runner (b) Riser
    (c) Pattern (d) Mould
51. _______ should be provided to avoid blow holes
    (a) Proper ventilation (b) Use gagger
    (c) Use strainer (d) Change gating design
52. Casting defect which produces a low projection on the drag surface of casting near to gate is
    (a) Dross (b) Blow holes
    (c) Wash (d) Gas holes
53. Defect caused due to presence of hydrogen in molten metal is
    (a) Wash (b) Pin holes
    (c) Dross (d) Drop
54. Which defect is caused due to high residual stress
    (a) Hot tear (b) Buckle
    (c) Scab (d) Scar
55. Which defect is caused due to slight compression fracture of thin layer of moulding sand
    (a) Scar (b) Rat tail
    (c) Cold shut (d) Blister
56. Defect caused due to misalignment between two halves of a mould is
    (a) Run out (b) Rat tail
    (c) Shift (d) Blister
57. In which die casting machine, furnace is integral with casting machine
    (a) Hot chamber (b) Cold chamber
    (c) Both (d) None of these
58. Hot chamber die casting machine is suitable for
    (a) High melting point alloys
    (b) Low melting point alloys
    (c) Medium melting point alloys
    (d) All of these
59. Which method is suitable for aluminium alloys and brass
    (a) Hot chamber die casting machine
    (b) Cold chamber die casting machine
    (c) Both (d) None of these
60. Function of tuyerers in cupola furnace is
    (a) To supply air (b) To remove slag
    (c) To feed charge (d) All of these
61. _______ is used to remove the oxides and impurities present in metal in form of slag
    (a) Charge (b) Coke
    (c) Flux (d) None of these
62. Which hole is used to remove the slag
    (a) Top hole (b) Slag hole
    (c) Tuyerers (d) Spout

63. Which zone is also known as superheating zone
(a) Oxidizing zone (b) Reducing zone
(c) Melting zone (d) Preheating zone

64. Temperature in oxidizing zone of cupola furnace is
(a) 1200 to 1400°C (b) 1530 to 1860°C
(c) 800 to 1200°C (d) 600 to 800°C

65. In which zone of cupolar furnace carbon dioxide reduces to carbon monoxide
(a) Melting zone (b) Preheating zone
(c) Oxidizing zone (d) Protective zone

66. Which zone lies above the protective zone and starts from first layer of metal charge above coke bed
(a) Melting zone (b) Preheating zone
(c) Oxidizing zone (d) Protective zone

67. Lower part of moulding flask is known as
(a) Cope (b) Drag
(c) Chaplet (d) Pattern

68. Cold shuts are found in
(a) Forming (b) Casting
(c) Welding (d) Machining

69. Molding sand is called green sand because it contains **[UPTU Dec. 2009]**
(a) Bakelite (b) Moisture
(c) No moisture (d) Dry sand

D. Match the Following

1.

(A)	(B)
(i) Pattern	(a) Taper on the pattern
(ii) Chills	(b) Produce the cavity in casting
(iii) Draft	(c) Replica of the product
(iv) Core	(d) Increase the cooling rate of casting.

2.

(A)	(B)
(i) Collapsibility	(a) Withstand high temp of molten metal
(ii) Refractoriness	(b) Allow gases to escape through the sand mould
(iii) Green strength	(c) Permit free contraction of metal
(iv) Permeability	(d) Retain the shape of mould

3.

(A)	(B)
(i) Non-metallic particles in metal	(a) Hot tear
(ii) Cracks due to high residual stress	(b) Inclusions
(iii) Produced due to dissociation of water	(c) Misruns
(iv) Unfilled cavities due to improper filling of mould cavity	(d) Pin hole

4.

	(A)		(B)
(i)	Hot chamber die casting machine	(a)	Regulate the molten metal
(ii)	Cold chamber die casting machine	(b)	Furnace is integral with casting
(iii)	Cupola furnace	(c)	Furnace is not integral with casting
(iv)	Runner	(d)	To melt cast iron

5.

	(A) Zone		(B) Temperature
(i)	Oxidizing zone	(a)	1600°C
(ii)	Reducing zone	(b)	1100°C
(iii)	Melting zone	(c)	1200°C
(iv)	Preheating zone	(d)	1530 to 1860°

ANSWERS

A. True or False

1. False 2. False 3. False 4. True 5. True

B. Fill in the Blanks

1. Replica
2. Core
3. Increase
4. Draft allowance
5. Fusing
6. Core
7. Gases
8. High residual stresses.

C. Multiple Choice Questions

1. (a)	**2.** (a)	**3.** (b)	**4.** (c)	**5.** (a)
6. (b)	**7.** (c)	**8.** (c)	**9.** (a)	**10.** (b)
11. (d)	**12.** (b)	**13.** (c)	**14.** (b)	**15.** (d)
16. (b)	**17.** (a)	**18.** (b)	**19.** (d)	**20.** (c)
21. (d)	**22.** (c)	**23.** (a)	**24.** (c)	**25.** (b)
26. (a)	**27.** (b)	**28.** (c)	**29.** (a)	**30.** (b)
31. (a)	**32.** (c)	**33.** (d)	**34.** (b)	**35.** (b)
36. (d)	**37.** (c)	**38.** (c)	**39.** (a)	**40.** (c)
41. (a)	**42.** (b)	**43.** (b)	**44.** (d)	**45.** (b)
46. (b)	**47.** (a)	**48.** (c)	**49.** (d)	**50.** (b)
51. (a)	**52.** (c)	**53.** (b)	**54.** (a)	**55.** (b)
56. (c)	**57.** (a)	**58.** (b)	**59.** (b)	**60.** (a)
61. (c)	**62.** (b)	**63.** (a)	**64.** (b)	**65.** (d)
66. (a)	**67.** (b)	**68.** (b)	**69.** (b)	

D. Match the Following

1. (i)—(c), (ii)—(d), (iii)—(a), (iv)—(b) **2.** (i)—(c), (ii)—(a), (iii)—(d), (iv)—(b)
3. (i)—(b), (ii)—(a), (iii)—(d), (iv)—(c) **4.** (i)—(b), (ii)—(c), (iii)—(d), (iv)—(a)
5. (i)—(d), (ii)—(c), (iii)—(a), (iv)—(b)

❑❑❑

UNIT-IV

Introduction to Machining and Welding and Its Applications

7. MACHINES AND MACHINING OPERATIONS

8. WELDING

SYLLABUS

Machining : Basic Principles of Lathe-Machine and Operations Performed on it. Basic Description of Machines and Operations of Shaper-Planer, Drilling, Milling and Grinding.

Welding : Importance and Basic Concepts of Welding, Classification of Welding Processes. Gas-Welding, Types of Flames. Electric-Arc Welding. Resistance Welding. Soldering and Brazing and its Uses.

CHAPTER 7

Machines and Machining Operations

INSIDE THIS CHAPTER

7.1. INTRODUCTION

Machining is a material removal process accomplished with the help of machine tool. A machine tool is a power driven device, which is used to impart a relative motion between the work piece and tool to perform various cutting operations. Various types of machine tools are used to produce different types of surfaces e.g. to produce cylindrical surface lathes are used and to produce flat surface shaper, planer, milling machine, grinding machine are used.

7.2. LATHE

It is one of the oldest, versatile and most widely used machine tool. Initially it was used for wood turning and with later developments it is made suitable for most cutting operations.

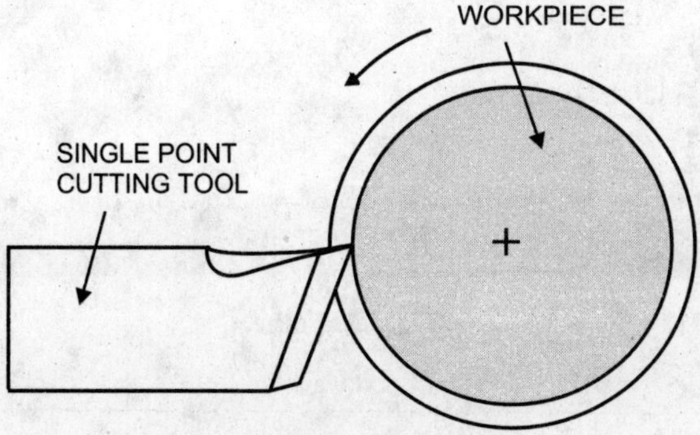

Fig. 7.1. Principle of Lathe.

Principle

Lathe removes the material in form of chips by rotating the workpiece against a single point cutting tool held in a tool post. Work piece is held between two support known as live centre and dead centre or in a chuck or face plate. Tool is fed perpendicular to axis of rotation of workpiece to decrease the diameter of workpiece and to generate flat surface. Either cylindrical or tapered surface are produced by lathe depending upon whether the tool moves parallel or at an inclination with the axis of rotation of workpiece.

7.3. PARTS OF A LATHE [UPTU DEC. 2008]

Main parts of lathe are as follows :

1. Bed
2. Headstock
3. Tailstock
4. Carriage
5. Feed Mechanism
6. Legs

Fig. 7.2. Lathe.

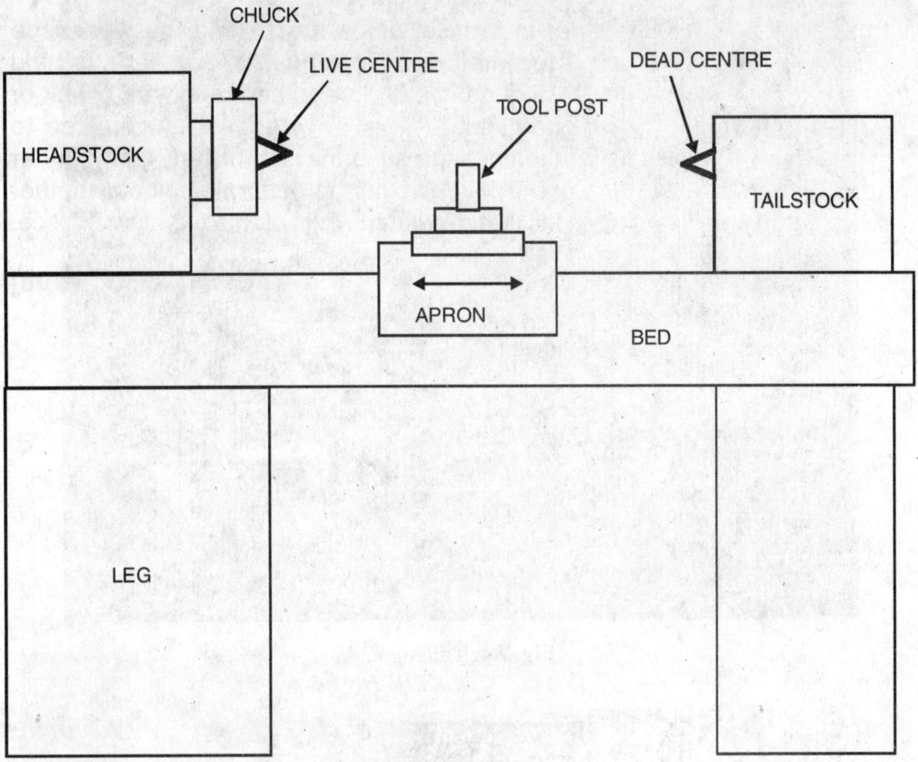

Fig. 7.3. Block Diagram of Lathe.

7.3.1. Bed

Bed of a lathe acts as the base on which, different parts are mounted and consist of two heavy metal slides having ways or V over it. Guide ways provide guided movement to carriage and tailstock. It is made up of cast iron due to its high damping capacity and good wear resistance.

7.3.2. Headstock

It is situated at the left hand of the bed and is fixed. It houses driving arrangement, transmission mechanism, headstock spindle, live centre etc. Headstock spindle is hollow through out its length and is supported on two bearings. Outer surface of spindle nose is generally threaded to mount the chuck or face plate.

7.3.3. Tailstock

Tailstock is a movable casting which can be moved along the ways to accommodate different length of workpiece. Upper part of the tailstock accommodates a hollow barrel whose one end is provided with a morse taper to hold dead centre or to support and feed tools to perform various operations like drilling, reaming etc.

Fig. 7.4. Tailstock.

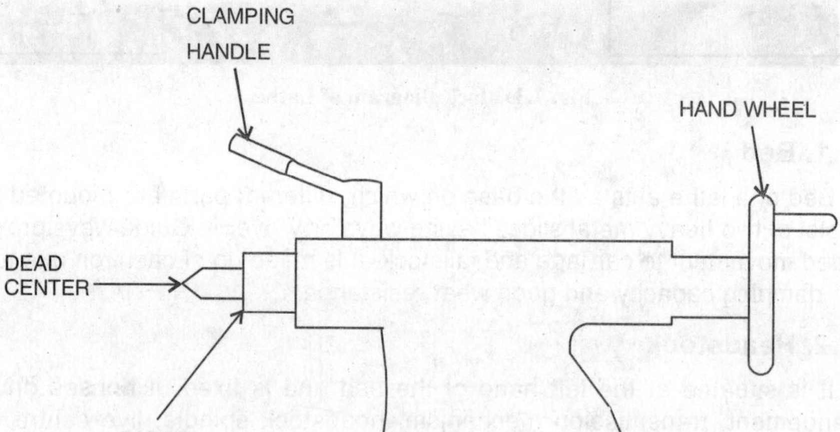

Fig. 7.5. Line diagram of tailstock.

7.3.4. Carriage

Carriage is a movable part located between headstock and tailstock. It is used for holding and feeding the tool against the job and has following main parts :

(i) Saddle

(ii) Cross Slide

(iii) Compound Rest

(iv) Tool Post

(v) Apron

Fig. 7.6. Carriage.

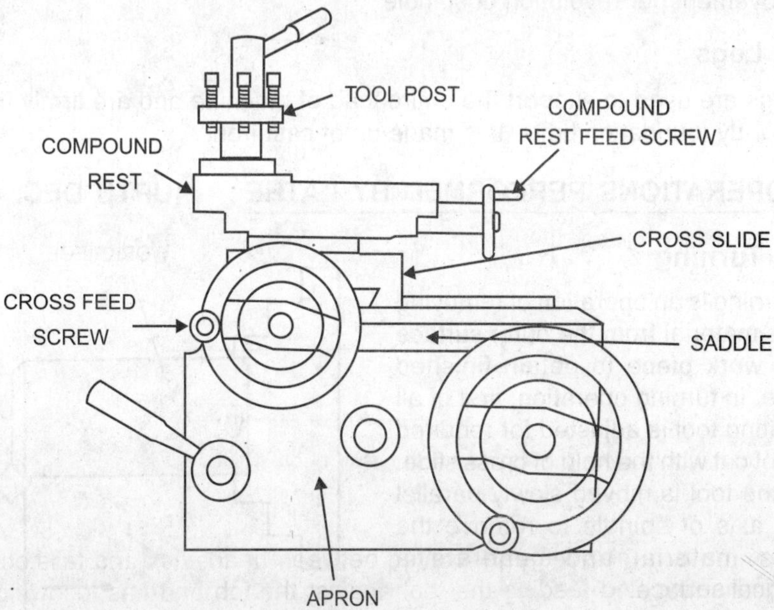

Fig. 7.7. Line diagram of carriage.

(i) Saddle

It is 'H' shape component that slide along the bed ways. It carries cross slide, compound rest and tool post.

(ii) Cross Slide

It is mounted on the top of the saddle to provide transverse tool movement and can be operated manually with a hand wheel or by power feed.

(iii) Compound Rest

It is mounted on the cross slide and has a circular base graduated in degrees. It can be swivelled about a vertical axis in various angular position. It is essential in most of taper turning operations.

(iv) Tool Post

It is mounted on the compound rest and is the top most part of carriage. It is used for holding various tools and tool holders.

(v) Apron

It is fastened to the saddle and hangs in front of the carriage. It consists of a split nut, gears and clutches to transmit motion from the feed rod to the carriage. Split nut is used to move the carriage in thread cutting operation.

7.3.5. Feed Mechanism

Feed mechanism is used to transmit power to the carriage. Initially power is transmitted by a gear train to quick change gearbox, which in turn regulates the tool movement per revolution of spindle.

7.3.6. Legs

Legs are used to support the entire load of machine and are firmly fixed to the floor by foundation bolts. It is made up of cast iron.

7.4. OPERATIONS PERFORMED BY LATHE [UPTU DEC. 2008]

7.4.1. Turning

Turning is an operation of removing excess material from the outer surface of the work piece to obtain finished surface. In turning operation, first of all the cutting tool is adjusted for required depth of cut with the help of cross slide. Then the tool is moved slowly parallel to the axis of spindle to remove the excess material and generate a cylindrical surface.

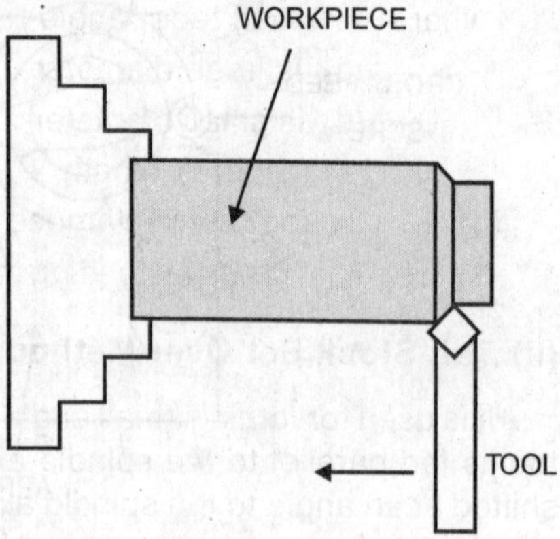

Fig. 7.8. Turning Operation.

7.4.2. Taper Turning [UPTU May 2009]

Tapered components are those, whose diameters are varying uniformly from one end to other end and taper is defined as the difference of diameters at two ends. Removing the excess material from the outer surface to generate tapered surface is known as taper turning. Various methods of taper turning are:

(i) Compound Rest Method
(ii) Tail Stock Set over Method
(iii) Using Form Tool Method
(iv) Taper Turning Attachment Method

(i) Compound Rest Method

It is used for turning short tapers with large angle. The step by step method is as follows :

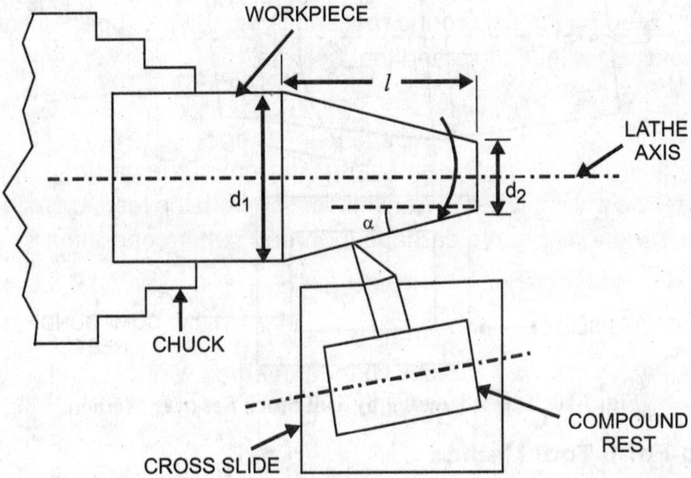

Fig. 7.9. Taper Turning by Compound Rest Method.

(a) In this method the compound rest is swivelled to the angle given by

$$\tan \alpha = \frac{d_1 - d_2}{2l}$$

where α is half taper angle
d_1 is large diameter
d_2 is small diameter
l is tapered length

(b) Compound rest is clamped in swivelled position.

(c) Taper is turned by feeding the tool by rotating the hand wheel of carriage.

(ii) Tail Stock Set Over Method

It is used for turning small angle of taper on long work pieces. In this method, tool is fed parallel to the spindle axis and the axis of rotation of work piece is shifted at an angle to the spindle axis by setting over the tailstock by a distance given by:

$$\text{Set over (S)} = L \times \frac{(d_1 - d_2)}{2l}$$

where S is set over

L is total length

l is length of tapered section

d_1 is large diameter

d_2 is small diameter

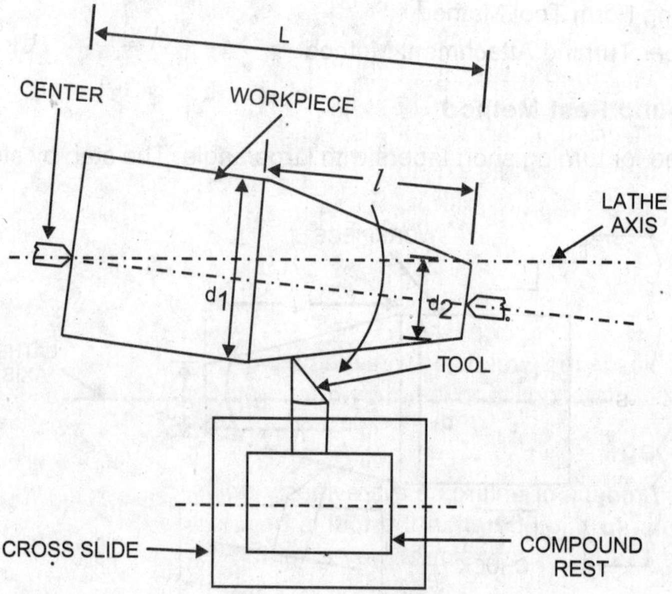

Fig. 7.10. Taper Turning by Tail Stock Set over Method.

(iii) Using Form Tool Method

Short external tapers can be turned with a broad nose form tool. The cutting edge of the tool is grounded at half the taper angle and is set straight into the workpiece to generate a taper.

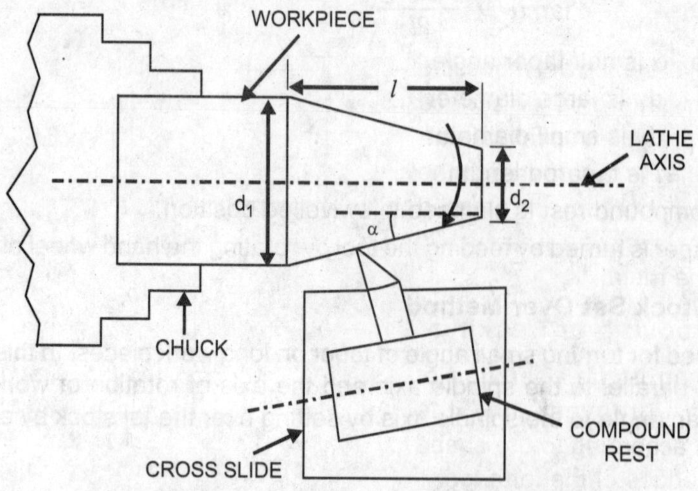

Fig. 7.11. Taper Turning by Using Form Tool Method

(iv) Taper Turning Attachment Method

This method is frequently used to produce long external tapers. It consists of guide block, bracket, guide bar, cross slide, etc. The attachment is fitted on the back of lathe and guide bar is set at half taper angle. As the carriage moves, a slide over the bar guides the tool according to guide bar and thus the taper is produced on the surface by keeping the centers in perfect alignment.

7.4.3. Facing

Facing is the operation of producing flat surface at the end of a work piece. The tool is feeded perpendicular to the spindle axis to obtain flat surface square with its own axis.

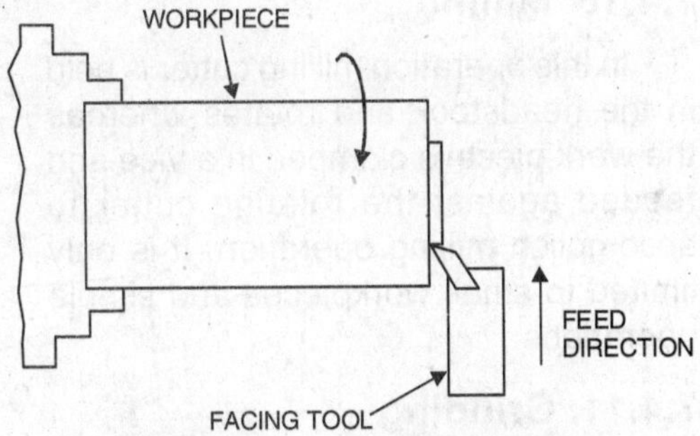

Fig. 7.12. Facing Operation.

7.4.4. Drilling

Drilling is the process of producing holes in the workpiece with the help of drills. For this operation, drill is held in the tail stock quill and is fed by rotating the hand wheel of the tailstock.

7.4.5. Boring

It is the process of enlarging a previously drilled hole produced by drilling, punching, etc. In this operation the tool is held in the tool post and is moved towards the headstock in a similar way as of turning.

7.4.6. Reaming

It is always carried out after drilling operation and is a process of finishing holes to accurate size and high dimensional accuracy. The reamer is held in the tail stock and is stationary while the workpiece revolves at a slow speed.

7.4.7. Knurling

It is an operation of embossing a particular pattern on the surface of work piece to provide an effective grip. In this operation the tool is held in tool post and pressed against the surface to plastically displace the metal.

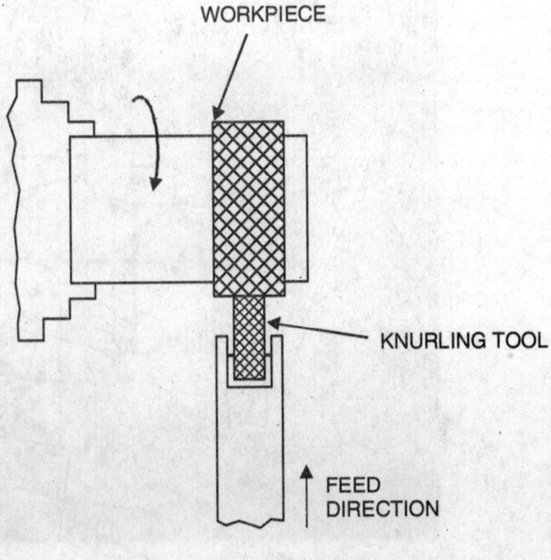

Fig. 7.13. Knurling Operation.

7.4.8. Grooving or Necking

Grooving is the operation of producing grooves on the workpiece according to the shape of cutting edges of the form tool.

7.4.9. Parting or Cutting Off

Parting is the operation of cutting away a required length of workpiece from a bar. It is done with a narrow cutting tool to reduce the wastage of material in the form of chips.

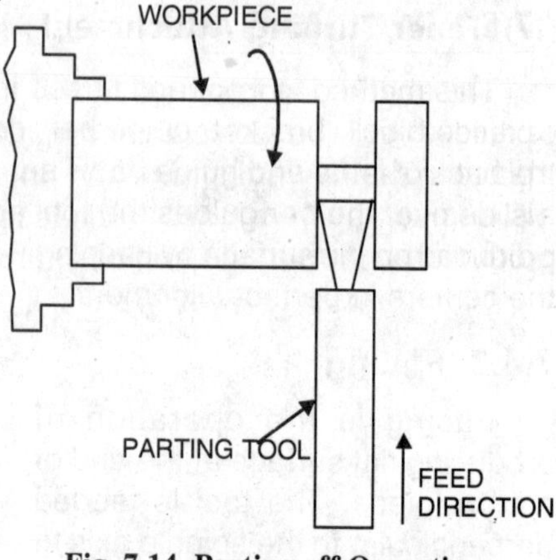

Fig. 7.14. Parting off Operation.

7.4.10. Milling

In this operation milling cutter is held in the headstock and rotates whereas the work piece is clamped in a vice and feeded against the rotating cutter to accomplish milling operation. It is only limited to small workpieces and simple operations.

7.4.11. Grinding

Internal as well as external grinding can be done on a lathe by using a tool post grinder. It is mounted on the compound rest and workpiece is held between the centers. Feed to the rotating grinding wheel is given by moving the carriage and depth of cut is adjusted by cross slide. It is used for simple operations such as sharpening of cutting tools, finishing surfaces of machined jobs, etc.

7.5. SHAPER [UPTU DEC. 2008, MAY 2009]

Shapers are intended to produce horizontal, vertical or inclined flat surfaces by means of a reciprocating single point cutting tool. Tool is given reciprocating motion by converting the circular motion into reciprocating motion with the help of mechanism.

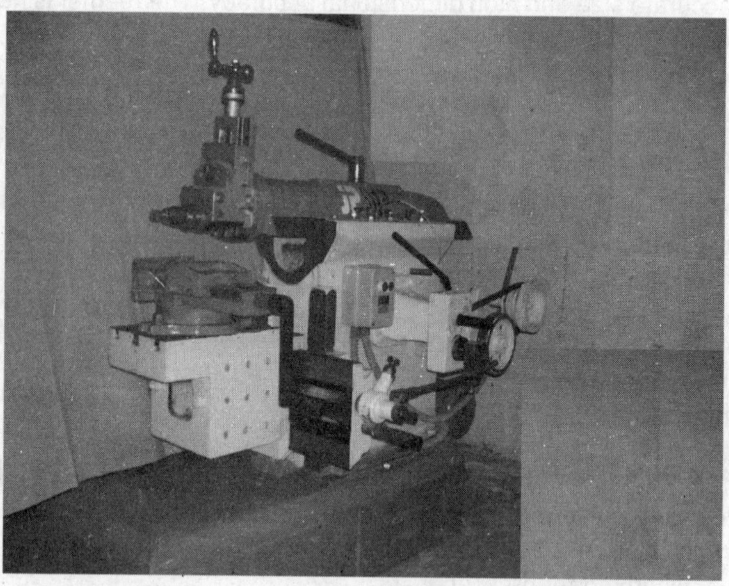

Fig. 7.15. Shaper.

7.5.1. WORKING PRINCIPLE

In shaper, tool is held in the tool post of reciprocating ram and the work piece is held in a vice clamped rigidly on the machine table. Cutting action takes place in the forward stroke and no material is removed in the return stroke so it is termed as idle stroke. Workpiece is given indexed feed perpendicular to the direction of ram at the end of return stroke.

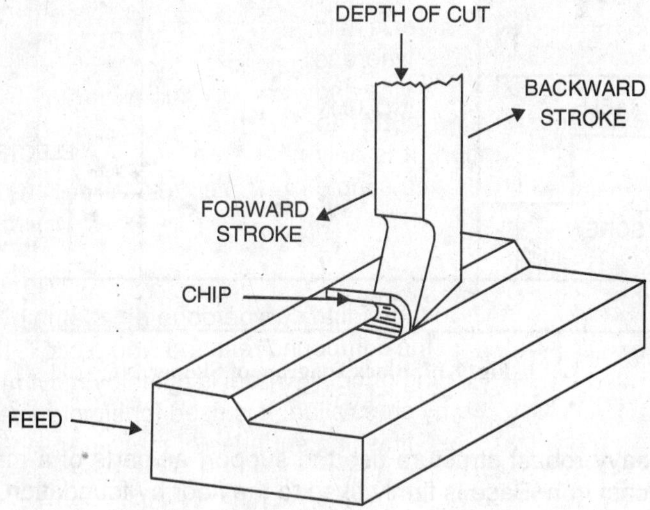

Fig. 7.16. Working Principle of Shaper.

Since the return stroke is idle, quick return mechanism is used in shaper to reduce the return idle stroke time.

7.5.2. CLASSIFICATION OF SHAPER

(i) According to the motion of ram

(a) Horizontal shaper
(b) Vertical shaper
(c) Travelling head shaper

(ii) According to mechanism driving the ram

(a) Crank type
(b) Geared type
(c) Hydraulic shaper

(iii) According to the movement of table

(a) Plain shaper
(b) Universal shaper

(iv) According to direction of cutting stroke

(a) Push cut
(b) Draw cut

7.5.3. Parts of a Shaper

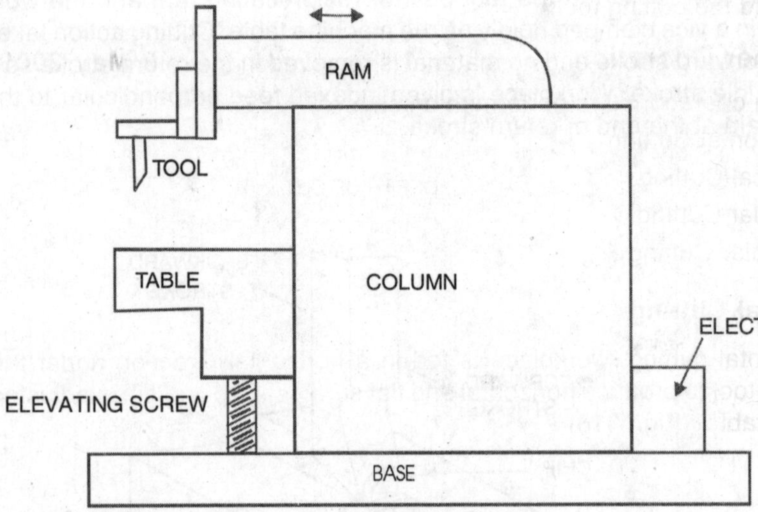

Fig. 7.17. Block Diagram of Shaper.

(i) Base

It is a heavy robust structure used to support all parts of a machine. It is made up of cast iron. Base is firmly fixed to the floor by foundation bolts.

(ii) Column

Column is a hollow casting, which is mounted on the base and acts as housing for drive mechanisms. On the top of the column there are guide ways for ram and at the front there are guide ways for cross rail.

(iii) Cross Rail

It is a heavy cast iron structure mounted on the front vertical guide ways of the column. It carries mechanisms for cross traverse and upward or downward motion of table.

(iv) Table

Shaper table are made of cast iron and have T-slots for holding the workpiece during the operation. Table moves along the cross rail to give feed to the workpiece.

(v) Ram

Ram carries the tool head at its front and moves in the guide ways above the column to impart a straight line motion to the tool. Tool head is mounted on the front of ram.

(vi) Tool Head

Tool head is fitted to the front end of ram on a circular plate. It can move upward, downward and also can be swivelled on circular plate for angular cuts. A clapper box is also provided on the tool head to prevent the dragging of tool on finished surface in return stroke. Clapper box raises the tool during the return stroke to prevent damage of tool cutting edge and machined surface. Tool holder is used to hold the cutting tool.

7.5.4. Shaper Operation [UPTU May 2009]

Following operations are performed on shaper

(i) Horizontal Cutting
(ii) Vertical Cutting
(iii) Angular Cutting
(iv) Irregular Cutting

(i) Horizontal Cutting

In horizontal cutting, workpiece is fed in a horizontal direction under the reciprocating tool to produce horizontal and flat surface. Depth of cut is adjusted by elevating table. (Fig. 7.18)

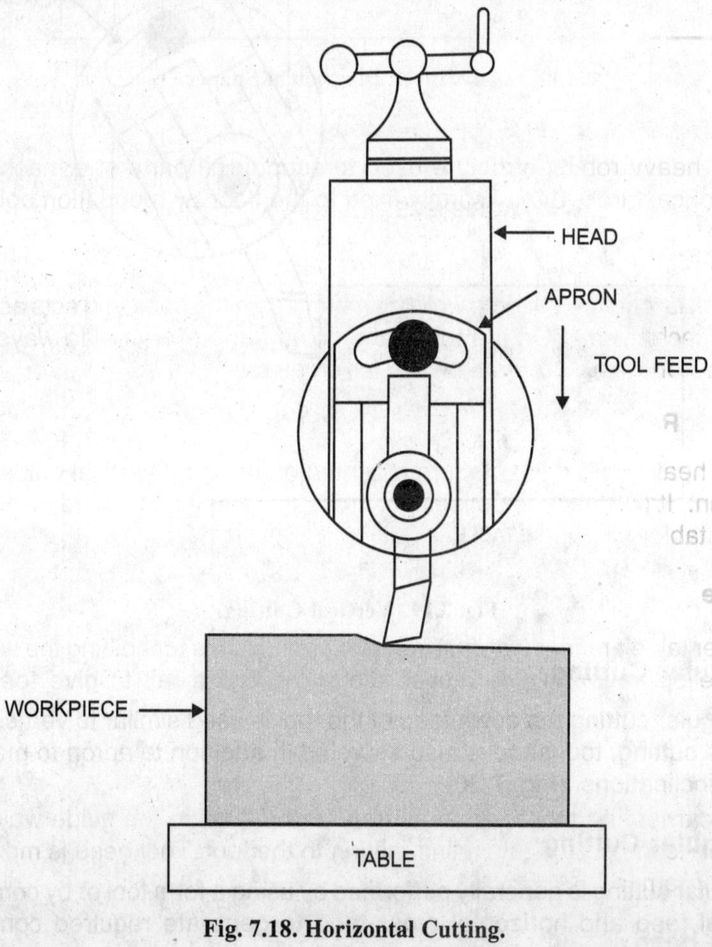

Fig. 7.18. Horizontal Cutting.

(ii) Vertical Cutting

It is used for squaring the ends and finishing the sides of workpiece. In this cutting, apron top is tilted except for parting off operations in a direction away from surface to be machined and then the tool is fed downward. (Fig. 7.19)

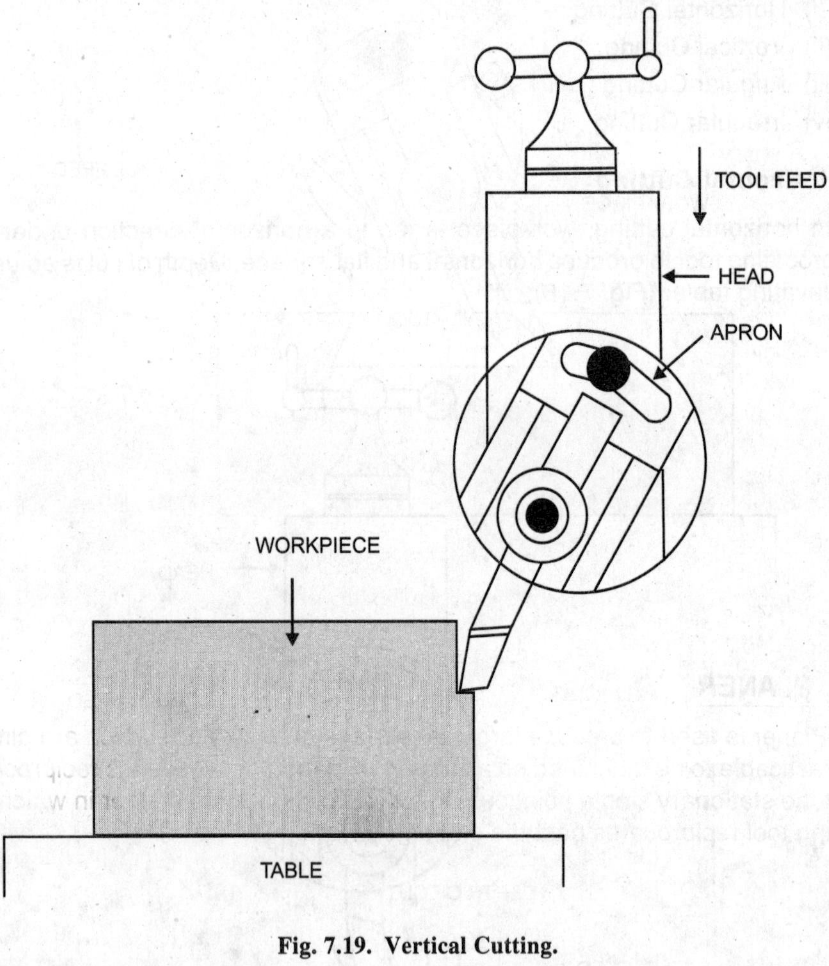

Fig. 7.19. Vertical Cutting.

(iii) Angular Cutting

In angular cutting the down feed of the tool is used similar to vertical cutting but in this cutting, tool head is also swiveled in addition to apron to machine at required inclinations. (Fig. 7.20)

(iv) Irregular Cutting

Irregular cutting is generally performed by using a form tool or by combination of vertical feed and horizontal cross-feed to generate required contours on surface.

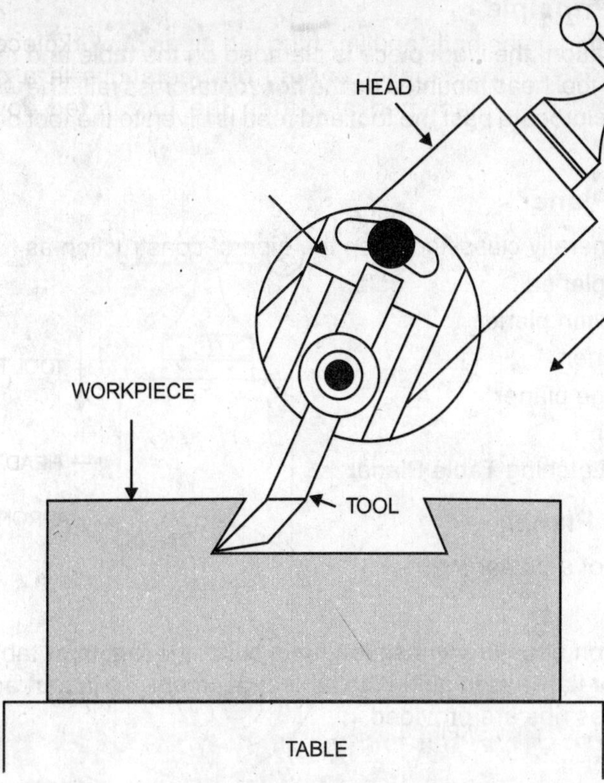

Fig. 7.20. Angular Cutting.

7.6. PLANER

Planer is used to produce large flat surface on workpiece which are almost impracticable to be machined on a shaper. In planer the workpiece reciprocates past the stationary single point cutting tool as compared to shaper in which the cutting tool reciprocates past the stationary work.

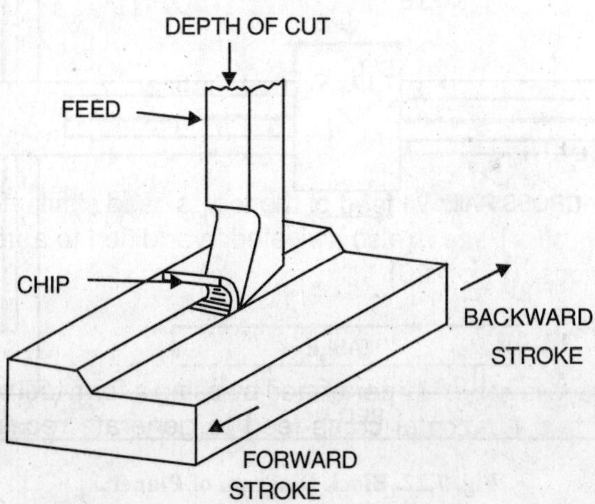

Fig. 7.21. Working Principle of Planer.

7.6.1. Working Principle

In planing operation, the work piece is clamped on the table and the tool is held vertically in the tool head mounted on the horizontal cross rail. The workpiece along with table, reciprocate past the tool and feed is given to the tool during the idle stroke.

7.6.2. Types of Planer

Planers are generally classified upon the type of construction as

(i) Open side planer
(ii) Double column planer
(iii) Pit type planer
(iv) Plate or edge planer
(v) Plane Miller
(vi) Divided or Latching Table Planer

7.6.3. Parts of a Planer

Principal parts of a planer are

(i) Bed

Bed is a cast iron structure whose length is twice the length of table. At its top, it has V ways or flatways to guide the table movement. To impart additional strength to bed cross ribs are provided.

(ii) Table

It is made of cast iron and moves over the bed ways. Its top surface has T slots to clamp the workpiece. It is driven by hydraulic cylinder or variable speed electric motor.

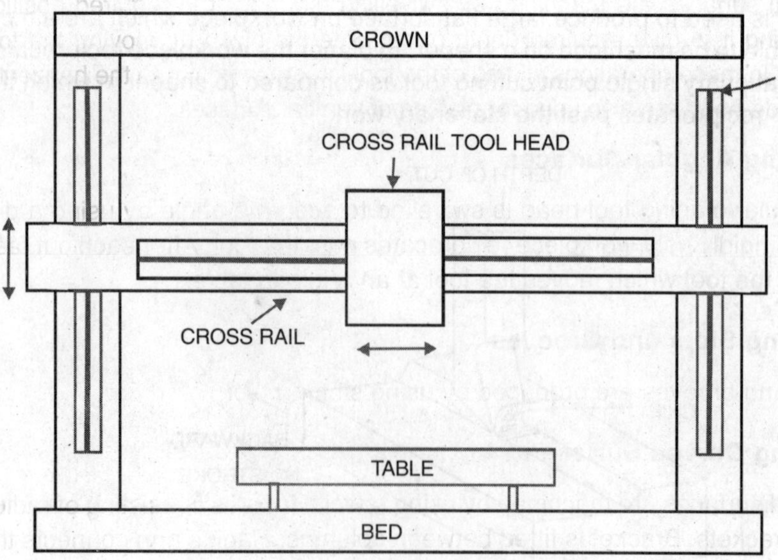

Fig. 7.22. Block Diagram of Planer.

(iii) Columns

Columns are vertical castings placed on both side of bed in double housing planer and on one side of bed in open side planer. Front surface of the column has guide ways for accurate movement of cross rails. Side tool head also moves along the columns.

(iv) Crossrail

It is a horizontal member connecting the two column. It is parallel to the table and supports the tool head. Cross-rail can be raised or lowered by means of elevating screws.

(v) Tool head

Tool head is mounted on saddle and saddle is mounted on the cross rail or columns. Tool head moves horizontally along the cross rail or vertically along the column and contains the tool post to hold the tool. Tool post is hinged to the tool head in such a way that the cutting tool is raised during the idle stroke. This saves damage of cutting edge of the tool.

7.6.4. Planer Operations

Common operations performed on a planer are as follows.

(i) Planing Horizontal Surfaces

For planing of horizontal surfaces, workpiece is held on table, proper tool is selected and crossrail is set to proper height. Workpiece along with the table reciprocates past the tool and feed is given to tool during idle stroke.

(ii) Planing Vertical Surfaces

Vertical surfaces are machined by swivelling tool holder in required position and adjusting it to give required depth of cut. Feed is given by moving the tool downward. In these operations sliding of the tool is perpendicular to the horizontal surface. Side-head are also used for planing of vertical surfaces.

(iii) Planing Angular Surfaces

In angular planing tool head is swivelled to accurate angle by using a dial gauge and rigidly held workpiece reciprocates past the tool. After each cut feed is given to the tool which moves the tool at an angle to table.

(iv) Planing Slots and Grooves

Slots and grooves are produced by using slotting tool.

(v) Planing Curved Surfaces

Curved surfaces are machined by using special fixtures consisting of radius arm and brackets. Bracket is fitted between columns. Radius arm connects the brackets and vertical slide of tool head. For concave planning, cross feed is engaged which causes the saddle to move along the crossrail and thus guides the tool to form concave surfaces.

7.6.5. Difference between Shaper and Planer

S.No.	Planer	Shaper
1.	It is heavy and robust in construction.	It is relatively lighter machine.
2.	Work piece reciprocates against the fixed tool.	Tool reciprocates against the fixed work-piece.
3.	It is meant for large work.	It can only do small work.
4.	Large depth of cut and coarse feed can be taken.	Large depth of cut and coarse feed is not possible due to overhanging ram.
5.	It can accommodate multiple tools.	It usually accommodates only one tool.
6.	Feed is given to tool during idle stroke of table.	Feed is given to the work during idle stroke of ram.

7.7. DRILLING

Drilling is an operation of producing circular holes in a workpiece by using a rotating tool known as drill. Drilling machines are most extensively used to produce holes.

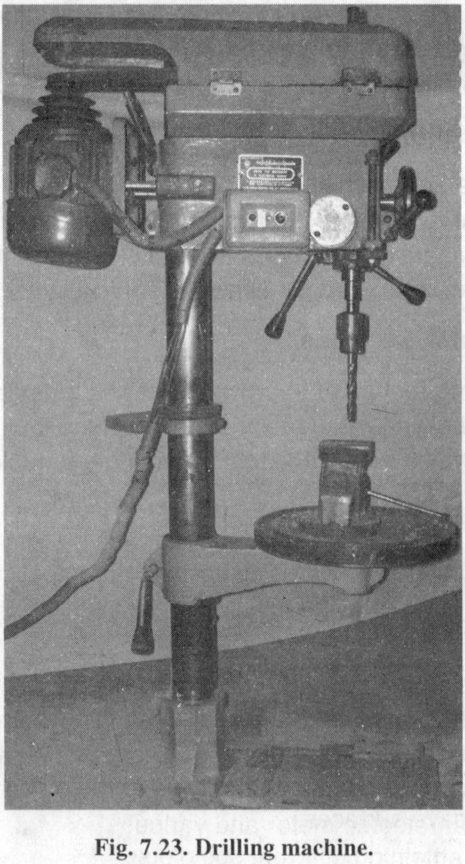

Fig. 7.23. Drilling machine.

Working Principle

In drilling operation, hole is produced by feeding the rotating drill in a direction parallel to its axis into a workpiece fixed to the table. During cutting the metal comes in contact with the two cutting edges of the tool and removes the metal stock. These cutting edges are known as lips of the drill.

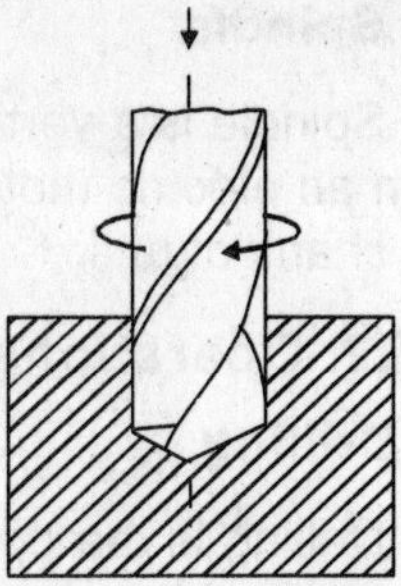

Fig. 7.24. Working Principle of drilling.

7.7.1. Parts of Drilling Machine

A drilling machine consist of

(i) Base

It is a large casting used to support the vertical column. In some machines base is provided with T slots to serve as table.

(ii) Column

Column is a cylindrical vertical member of the machine used to support the table and head.

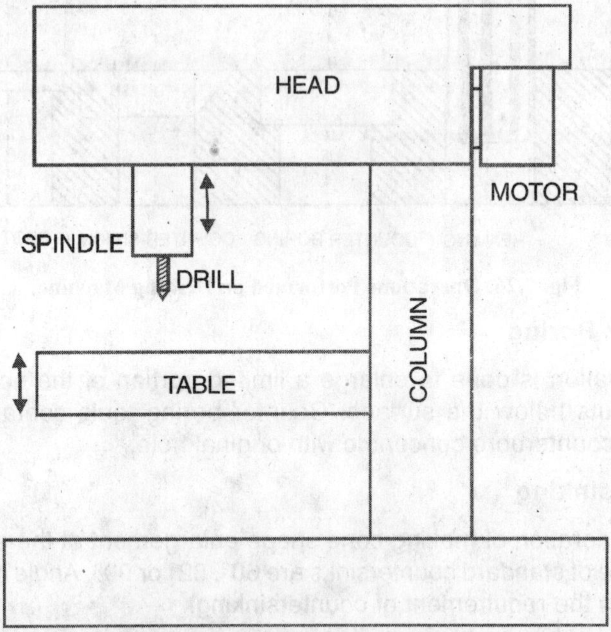

Fig. 7.25. Block Diagram of Drilling Machine.

(iii) Table

It is mounted on column and is used to clamp the workpiece. It may be rectangular or circular in shape depending upon the requirement of machine.

(iv) Head

Head consists of an electric motor and various devices such as step cone pulley, gears, etc. for obtaining different spindle speeds.

(v) Spindle

Spindle is a vertical shaft which rotates on bearings by the power received from an electric motor. Its lower end is connected to a drill chuck which holds the shank of drill.

7.7.2. Operations Performed on Drilling Machine

(i) Drilling

Drilling is the operation of producing circular holes in a workpiece using a rotating tool known as drill.

(ii) Reaming

It is performed by inserting the reamer into the spindle and rotating it in a previously drilled hole to bring it in accurate size.

(iii) Boring

It is performed by a boring bar having a single cutting edge to enlarge a previously drilled hole. It also corrects the position of hole which is not possible by reaming operation.

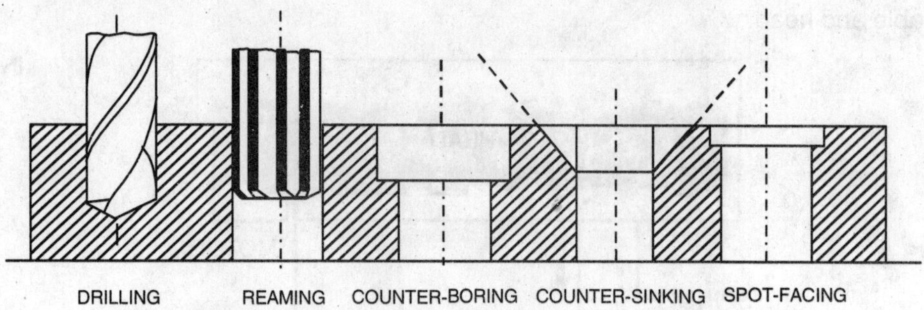

Fig. 7.26. Operations Performed on Drilling Machine.

(iv) Counter Boring

This operation is done to enlarge a limited portion of the hole to set bolt heads and nuts below the surface. Counter boring tools contains a pilot to maintain the counterbore concentric with original hole.

(v) Countersinking

It is the operation of making cone shape enlargement at the end of a hole. Included angle of standard countersinks are 60°, 82° or 90°. Angle of countersink depends upon the requirement of countersinking.

(vi) Tapping

Tapping is an operation meant for producing internal threads in a drilled hole by using a tool known as tap. Tap must be perfectly aligned with the centre axis of hole to avoid its breaking.

(vii) Spot Facing

This operation is performed for squaring the surface around a drilled hole to provide a seat for bolt head, collar, nut, washer, etc. Spot facing is usually performed on forgings and castings.

7.8. MILLING

Milling is the machining operation in which material is removed from the surface by using a multipoint rotating cutting tool known as milling cutter. Milling cutter is known as multipoint cutting tool because it has equally spaced cutting teeth either on periphery or on end or on both. This machine is most widely used due to fast production, high accuracy and good surface finish.

Fig. 7.27. Milling machine.

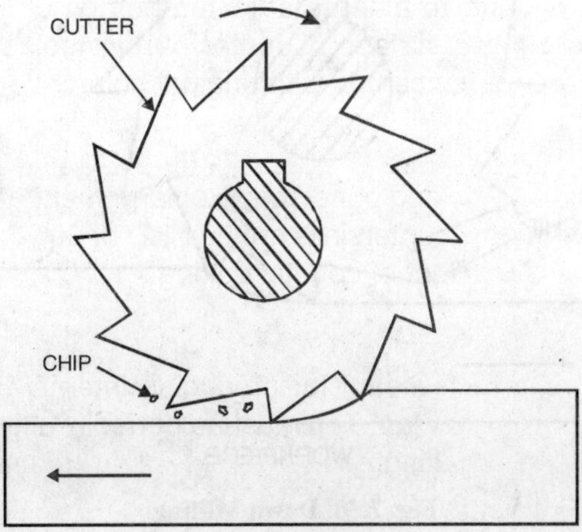

Fig. 7.28. Working Principle of Milling.

7.8.1. Working Principle [UPTU May 2009]

In milling operation, workpiece is held rigidly on the table and the milling cutter is mounted on a spindle or on arbor. For cutting operation the workpiece is fed slowly past a rotating cutter. The cutter rotates at a very high speed and removes the material in the form of chips to produce desired shape (Fig. 7.28). Workpiece can be fed in a vertical, longitudinal or cross direction as per requirement.

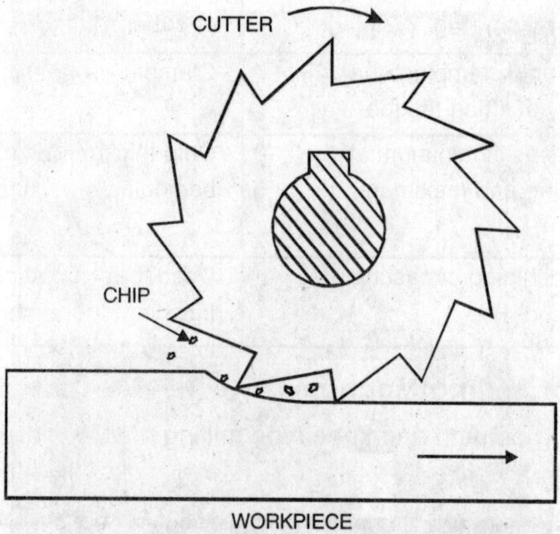

Fig. 7.29. Up Milling.

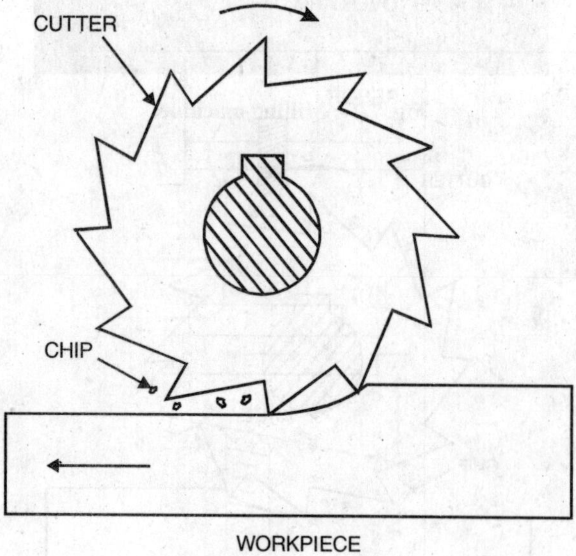

Fig. 7.30. Down Milling.

7.8.2. Up Milling and Down Milling [UPTU May 2009]

Comparison between up milling and down milling.

S.No.	Up Milling	Down Milling
1.	It is also known as conventional milling.	It is also known as climb milling.
2.	Work piece is fed in direction opposite to cutter rotation. (Fig. 7.29)	Work piece is fed in the direction of cutter rotation. (Fig. 7.30)
3.	It requires rigid clamping due to tendency of lifting the job.	Clamping is not so crucial.
4.	Chip thickness is minimum at the beginning and maximum at the end.	Chip thickness is maximum at beginning and minimum at the end.
5.	Used in machining of castings and forgings.	Used for slot cutting, surface finishing operation etc.

7.8.3. Parts of Milling Machine

Main parts of a column and knee type milling machine are as follows :

(i) Base

It is a cast iron casting on which column is mounted and is used to support all other parts. It also carries screw jack to move the knee.

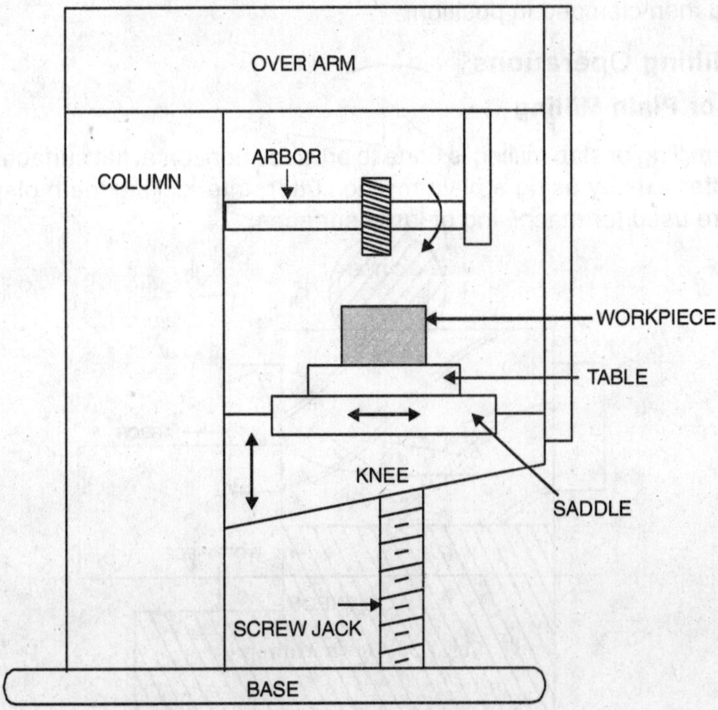

Fig. 7.31. Block Diagram of Milling Machine.

(ii) Column

It encloses electric motor, driving mechanism, etc. and supports all other parts of machine. At its front it contains the guide ways to move the knee up and down.

(iii) Knee

Knee supports saddle and table and is capable of moving up and down on the guide ways of column. Its top surface contains horizontal ways for cross traverse of saddle. It also encloses the gear mechanisms to change the feed.

(iv) Saddle

It is the part between knee and table and is used to provide cross-feed to table. On top of the saddle there are horizontal ways for longitudinal traverse of table.

(v) Table

Table moves over the guide ways of the saddle to impart longitudinal feed. Top surface of table contains T-slot to rigidly clamp the workpiece with clamping bolts.

(vi) Spindle

It is a hollow shaft that rotates in bearing and receives power from electric motor through a pulley and belt or gear arrangement.

(vii) Arbor

Arbor is an accurately machined shaft on which milling cutters are mounted. Its one end is tapered to fit the spindle nose and has two slots to fit the nose keys for locating and driving it.

(viii) Overarm

Overarm is used for supporting arbors and is mounted on the top of the column. It is moved in and out by hand to the position of maximum support for arbor and then clamped in position.

7.8.4. Milling Operations

(i) Slab or Plain Milling

Plain milling or slab milling is done to produce horizontal flat surfaces parallel to the cutter axis by using a plain milling cutter. Interlocking teeth plain milling cutters are used for machining of large surfaces.

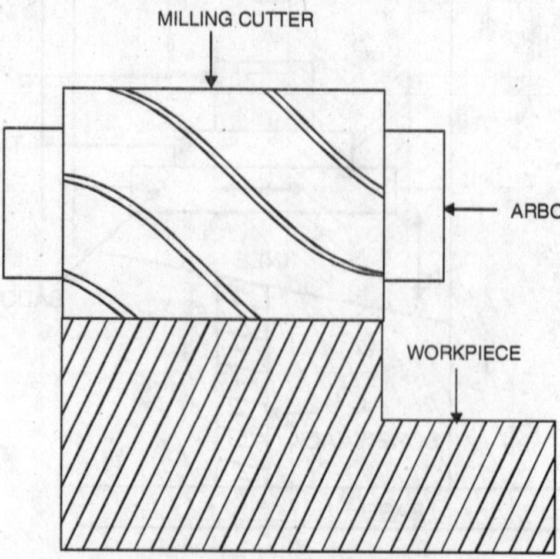

Fig. 7.32. Plain Milling.

(ii) Face Milling

Face milling is done to produce flat surfaces perpendicular to the axis of rotation of face milling cutter. In face milling, the cutter is mounted on a spindle whose axis of rotation is perpendicular to the work surface.

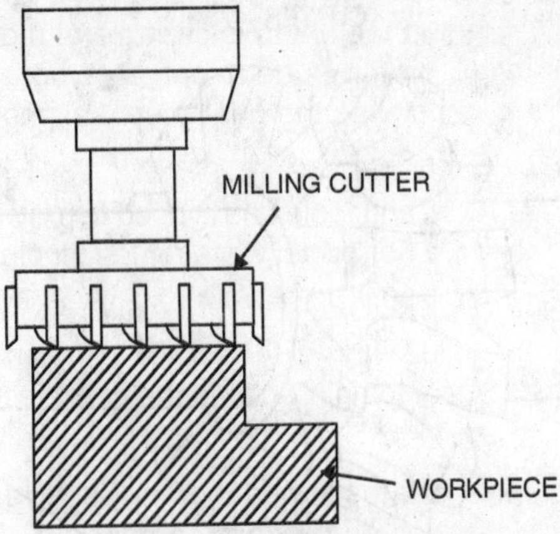

Fig. 7.33. Face Milling.

(iii) Angular Milling

Angular milling is the operation of producing flat surfaces at any angle other than right angle to the axis of cutter. Two inclined surfaces can be machined simultaneously by using a double angle cutter.

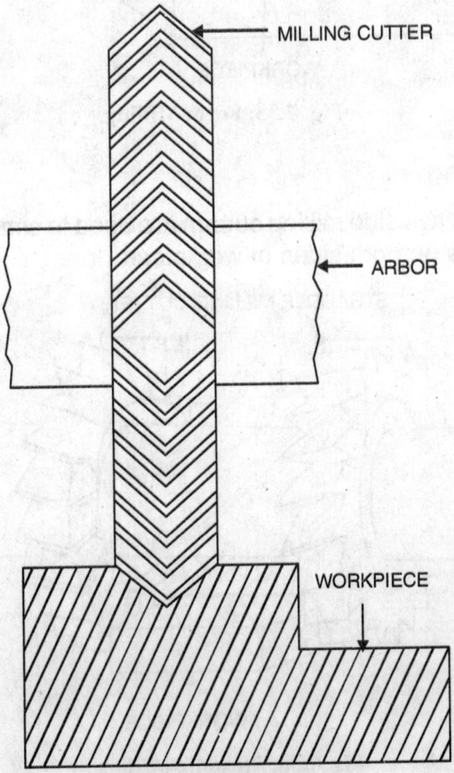

Fig. 7.34. Angular Milling.

(iv) Form Milling

In form milling irregular counters are produced by using form cutters. These cutters are known as form milling cutters because their teeth have profile according to the surface to be generated.

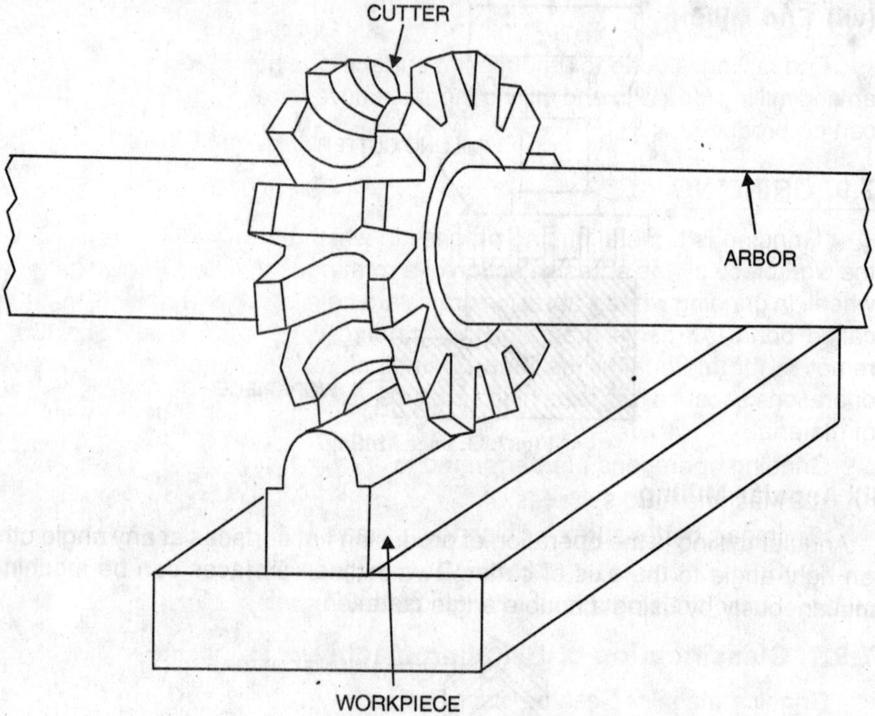

Fig. 7.35. Form Milling.

(v) Straddle Milling

In this operation two side milling cutters are used to simultaneously machine flat vertical surfaces on both sides of workpiece.

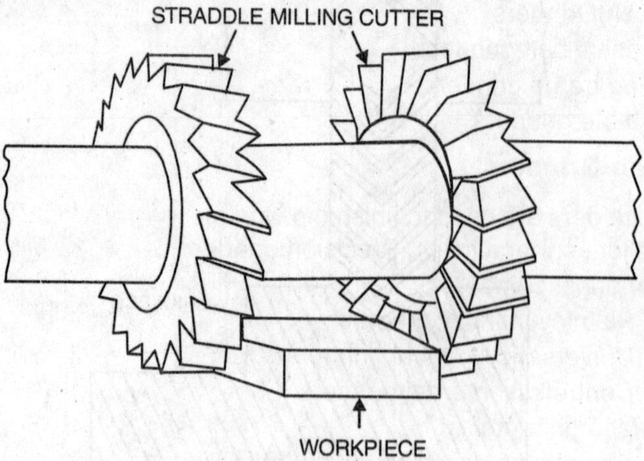

Fig. 7.36. Straddle Milling.

(vi) Gang Milling

If more than two cutters are mounted on the same arbor to machine a number of surfaces simultaneously, than this type of milling is known as gang milling. It is used to reduce the machining time by performing simultaneous operations.

(vii) End Milling

End milling is done to produce flat surface or a pair of flat surfaces by using an end milling cutter. In end milling, horizontal, vertical as well as inclined surface can be produced.

7.9. GRINDING

Grinding is a metal cutting process in which the material is removed from the workpiece by the abrasive action of a rotating cutting tool known as grinding wheel. In grinding wheels abrasive grains are held together by a binding material, called bond. Abrasive grains of the grinding wheel acts as cutting tool and removes the material in the form of small chips. Grinding is used in finishing operations because it imparts high surface finish by removing a very small amount of material.

Grinding operations are performed to

1. Machine hard surfaces.
2. Impart better surface finish to surfaces.
3. Sharpen the cutting tools.
4. Finish the work piece within close tolerances, etc.

7.9.1. Classification of Grinding Machines

Grinding machines can be classified as

1. Rough Grinders
2. Precision Grinders

1. Rough Grinders

These grinders are used for removal of material rather than for accuracy. Most commonly used rough grinders are

(a) Bench Grinders
(b) Abrasive belt grinders
(c) Swing frame grinders
(d) Portable grinders

2. Precision Grinders

These grinders are used to finish the workpiece within close tolerances and impart excellent surface finish. Precision grinders can be further classified as

(a) Cylindrical Grinders
 (i) Plain cylindrical grinders
 (ii) Universal cylindrical grinders
 (iii) Centreless internal grinders
(b) Surface grinders
 (i) Planer type or reciprocating table
  - Horizontal spindle
  - Vertical spindle

(ii) Rotating table
- Horizontal spindle
- Vertical spindle

(c) Internal grinders
(i) Plain grinders
(ii) Chucking grinders
(iii) Universal grinders
(iv) Centreless grinders
(v) Planetary grinders

(d) Tool grinders

(e) Special grinders

7.9.2. CYLINDRICAL GRINDERS

These are used to generate accurate surface on circular workpiece. It includes the holding of workpiece rigidly between centres, rotating it about its axis and feeding a rotating grinding wheel against the workpiece. Cylindrical grinding machines are

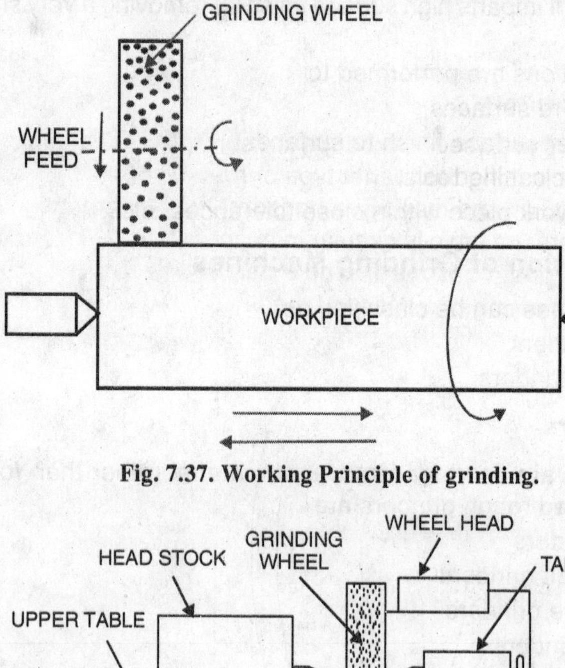

Fig. 7.37. Working Principle of grinding.

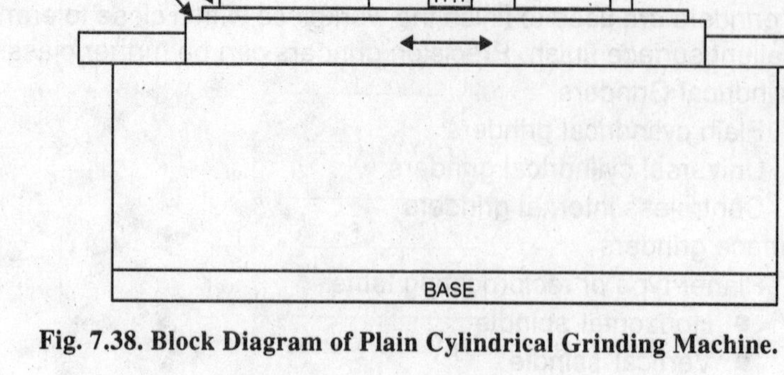

Fig. 7.38. Block Diagram of Plain Cylindrical Grinding Machine.

(i) Plain Cylindrical Grinders

In these grinders work piece is held between the headstock centre and the tailstock centre, and the rotating work is fed against the rotating grinding wheel (Fig. 7.38). Grinding wheel is mounted on the spindle of wheel head and travel along horizontal crossways. Tailstock moves longitudinally and can be clamped in any position to accommodate different length of workpiece.

(ii) Universal Cylindrical Grinders

Universal cylindrical grinders are similar to plain grinders with the following additional features for greater versatility :

(a) Live or dead head stock spindle

(b) Headstock swivel mechanism

(c) Wheel head swivel mechanism

(iii) Centreless Grinders

In these machines workpiece is supported by work-rest, grinding wheel and regulating wheel instead of centres to grind work piece which do not have centres such as piston, valves, etc.

7.9.3. Surface Grinders

Surface grinders are used to produce flat surfaces. According to the type of table these can be classified as planer type or rotary type. In planar type grinder the table is rectangular and may have horizontal or vertical spindles. In rotary type surface grinders the table is circular in shape and has horizontal or vertical spindles. In both types, horizontal spindle carries the straight wheel whereas vertical spindle carries the cup shaped wheel.

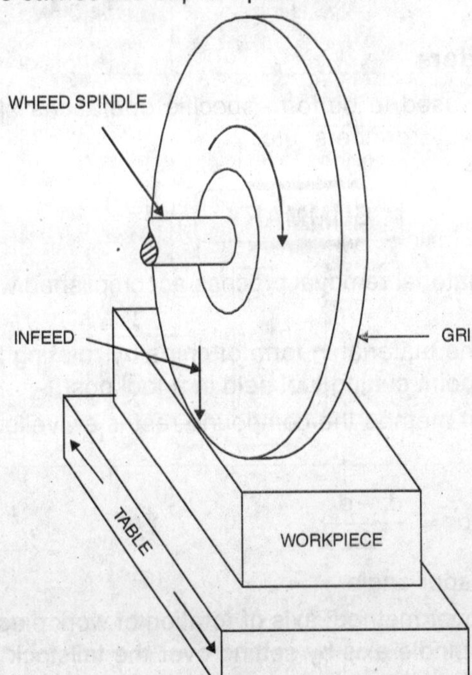

Fig. 7.39. Horizontal Spindle Reciprocating Table Surface Grinder.

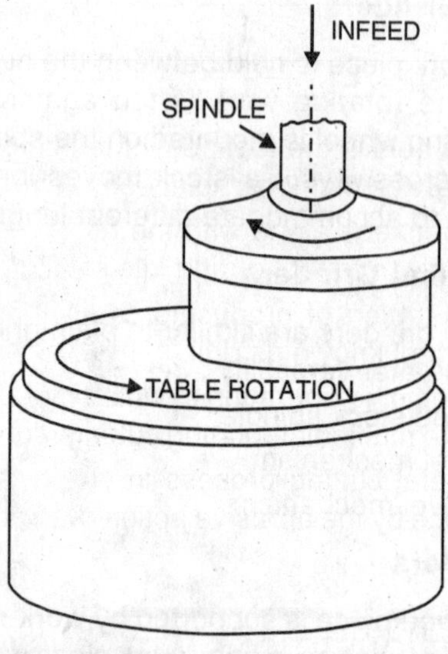

Fig. 7.40. Vertical Spindle Rotary Table Surface Grinder.

7.9.4. Internal Grinding Machine

These machines are used for grinding internal holes, tapers, cylindrical surface, etc. and are specially designed for the production of interchangeable parts with exact dimension and excellent surface finish.

7.9.5. Tool Grinders

These grinders are used to resharpen the cutting tools such as reamers, drills, milling cutters, etc.

7.9.6. Special Grinders

These grinders are used to perform specific operations such as thread grinders, cam grinders, ways grinders, etc.

SUMMARY

- Machining is a material removal process accomplished with the help of machine tool.
- Lathe removes the material in form of chips by rotating the workpiece against a single point cutting tool held in a tool post.
- In compound rest method the compound rest is swivelled to the angle given by

$$\tan \alpha = \frac{d_1 - d_2}{2l}$$

where α is half taper angle

- In tail stock set over method, axis of rotation of work piece is shifted at an angle to the spindle axis by setting over the tailstock by a distance given by

$$\text{Set over (S)} = L \times \frac{(d_1 - d_2)}{2l}$$

- In shaper, tool is held in the tool post of reciprocating ram and the work piece is held in a vice clamped rigidly on the machine table.
- In planer the workpiece along with the table reciprocate past the fixed tool and feed is given to tool during idle stroke.
- Drilling is an operation of producing circular holes in a workpiece by using a rotating tool known as drill.
- Milling is the machining operation in which material is removed from the surface by using a multipoint rotating cutting tool known as milling cutter.
- Grinding is a metal cutting process in which the material is removed from the workpiece by the abrasive action of a rotating cutting tool known as grinding wheel.

SHORT QUESTIONS WITH ANSWERS

Q. 1. What is Chuck ?

Ans. Chuck is a device used for holding the workpiece during the operation and is mounted on the spindle.

Q. 2. Differentiate between four jaw chuck and three jaw chuck.

Ans. In four jaw chuck each jaw moves independently of others and is used for exact centering whereas in three jaw chuck all the three jaws move together to hold the workpiece in centre but use of three jaw chuck is limited to round or hexagonal workpiece.

Q. 3. Determine the half taper angle for a job whose large diameter is 35 mm, small diameter is 15 mm and is tapered to a length of 25 mm.

Ans.
$$\tan\alpha = \frac{d_1 - d_2}{2l} = \frac{35-15}{2\times 25} = \frac{20}{50} = 21.8^\circ$$

Q. 4. Determine tailstock set over for a job whose length is 300 mm, large diameter is 50 mm, small diameter is 30 mm and tapered length is 200 mm.

Ans.
$$\text{Set over (S)} = \frac{L\times(d_1 - d_2)}{2\times l}$$

$$= \frac{300\times(50-30)}{2\times 200} = \frac{300\times 20}{400} = 15\text{ mm}$$

Q. 5. What is the working principle of shaper?

Ans. In shaper, tool is held in the tool post of reciprocating ram and the workpiece is held in a vice clamped rigidly on the machine table.

Cutting action takes place in forward stroke and indexed feed is given to workpiece perpendicular to the direction of ram at the end of return stroke.

Q. 6. Why clapper box is used in shaper?

Ans. Clapper box is used to prevent the dragging of tool on finished surface in return stroke by raising the tool, thus preventing the damage of cutting edge.

Q. 7. Which tool is used in milling?

Ans. Multi point cutting tool known as milling cutter is used in milling machine.

Q. 8. Why rigid clamping is required in up milling?

Ans. Up milling requires rigid clamping because it has a tendency to lift the job during machining.

Q. 9. What is the need of centerless grinders?

Ans. These are used to grind workpiece which do not have centers.

Q.10. Define grinding.

Ans. Grinding is a metal cutting process in which the material is removed from the workpiece by abrasive action of a rotating grinding wheel.

Q.11. What do you mean by gang milling. **[UPTU May 2012]**

Ans. In gang milling more than two cutter are mounted on same arbor to machine number of surfaces simultaneously.

Q.12. List the operations performed on lathe machine. **[UPTU May 2012]**

Ans. (i) Facing (ii) Drilling (iii) Taper turning (iv) Reaming

Q.13. How lathe size is specified. **[UPTU Dec. 2011]**

Ans. Lathe size is specified by any one of following ways

(i) Swing or maximum diameter that can be rotated over the bed ways.

(ii) Diameter of hole through lathe spindle for turning bar material.

(iii) Height of centres measured over lathe bed.

REVIEW QUESTIONS

1. Describe the working principle of lathe.
2. Explain various parts of a lathe with neat sketch. **[UPTU Dec. 2008; May 2010]**
3. Explain carriage with its main parts.
4. Explain various methods of taper turning.
5. Enlist various operations performed on lathe. **[UPTU Dec. 2008; May 2010]**
6. Give the classification of shaper.
7. Explain various parts of a shaper.
8. Differentiate between shaper and planer.
9. With the help of schematic diagram, describe the working principle and important parts of drilling machine. Also describe drilling operation. **[UPTU Dec. 2010]**
10. How drilling machines are classified?
11. Explain various parts of a drilling machine.
12. Define the milling process and differentiate between up milling and down milling.
13. What do you mean by grinding? **[UPTU May 2012]**
14. Give the classification of grinding machines.
15. Explain different types of cylindrical grinding machines.
16. With the help of schematic sketch, describe the basic working principle and important parts of shaper machine. Also describe the difference between shaper and planer. **[UPTU Dec. 2008]**
17. With the help of schematic sketch. Describe the basic working principle of grinding process. **[UPTU Dec. 2009]**
18. Explain the working principle and operation of milling machine with neat diagram. **[UPTU Dec. 2010]**

SELF TEST QUESTIONS

A. State Whether True or False

1. Headstock is a movable casting at the left hand of bed.
2. Apron is fastened to tail stock.
3. Cross slide is used to provide transverse tool movement.
4. Boring is used to produce hole in the workpiece.
5. In up milling, workpiece is fed in direction opposite to cutter rotation.
6. Down milling requires rigid clamping due to tendency of lifting the job.
7. In straddle milling two side milling cutters are used to machine vertical surface on both side of workpiece.
8. In centerless grinding, workpiece is supported between headstock centre and tailstock centre.
9. Shaper is used for machining of large flat surface.
10. Reaming is an operation of producing holes of accurate size and high dimensional accuracy.

B. Multiple Choice Questions

1. To produce a tapered surface on lathe tool must move
   (a) at an inclination with workpiece axis.
   (b) parallel to workpiece axis.
   (c) perpendicular to workpiece axis.
   (d) all of the above.
2. In compound rest method, compound rest is swivelled by the angle
   (a) $\tan\alpha = \dfrac{d_1 - d_2}{l}$ (b) $\tan\alpha = \dfrac{d_1 - d_2}{2l}$
   (c) $\tan\alpha = L \times \dfrac{d_1 - d_2}{l}$ (d) $\tan\alpha = L \times \dfrac{d_1 - d_2}{2l}$
   [UPTU Dec. 2009]
3. Set over in tail stock set over method is given by
   (a) $S = l \times \dfrac{(d_1 - d_2)}{L}$ (b) $S = l \times \dfrac{(d_1 - d_2)}{2L}$
   (c) $S = L \times \dfrac{(d_1 - d_2)}{l}$ (d) $S = L \times \dfrac{(d_1 - d_2)}{2l}$
4. In shaper
   (a) Workpiece reciprocates past a stationary tool.
   (b) Tool reciprocates past a stationary workpiece.
   (c) Tool and workpiece both reciprocates.
   (d) Tool rotates with respect to workpiece.
5. In planer
   (a) Workpiece reciprocates past a stationary tool.
   (b) Tool reciprocates past a stationary workpiece.
   (c) Tool and workpiece both reciprocates.
   (d) Tool rotates with respect to workpiece.

6. Quick return mechanism is used in shaper
   (a) To reduce cutting stroke time.
   (b) To reduce idle return stroke time.
   (c) For fast movement of table.
   (d) For fast cutting.
7. Milling is the machining operation in which cutting takes place by
   (a) Single point rotating tool.
   (b) Multi-point rotating tool.
   (c) Abrasive action of a rotating cutting tool.
   (d) None.
8. Grinding is the machining operation in which cutting takes place by
   (a) Single point cutting tool.
   (b) Single point reciprocating tool.
   (c) Abrasive action of a rotating cutting tool.
   (d) None
9. In plane cylindrical grinders, work is held between
   (a) Chuck and tailstock.
   (b) Chuck an grinding wheel.
   (c) Headstock centre and tailstock centre.
   (d) Jaw of vice.
10. Surface grinders are used to produce
    (a) Flat surfaces (b) Circular surfaces.
    (c) Irregular surfaces (d) None
11. Finishing is the operation best done by
    (a) Milling machine (b) Shaper machine
    (c) Planer machine (d) Grinding
12. Swing is one of the specification of **[UPTU May 2009]**
    (a) Lathe machine (b) Shaper machine
    (c) Drilling machine (d) Grinding machine
13. Cylindrical surfaces are produced by
    (a) Lathe (b) Shaper
    (c) Planer (d) Grinding
14. In lathe
    (a) Workpiece rotates (b) Tool rotates
    (c) Workpiece reciprocates (d) None of these
15. To decrease the diameter of workpiece on lathe, tool is fed
    (a) Perpendicular to axis of rotation
    (b) Parallel to axis of rotation
    (c) In zig-zag
    (d) In circular manner
16. When the tool moves at an inclination with the axis of rotation of workpiece, the surface produced is
    (a) Cylindrical surface (b) Tapered surface
    (c) Threaded surface (d) Rough surface
17. In lathe, workpiece is held between two supports known as
    (a) Bed and legs (b) Live centre and dead centre
    (c) Carriage and tailstock (d) Bed and live centre

18. Which is not a part of lathe
(a) Feed mechanism (b) Ram
(c) Head stock (d) Legs

19. Lathe bed is made up of
(a) Aluminium (b) Copper
(c) Cast-iron (d) Tungsten

20. Cast iron is used as material for lathe bed due to its
(a) High wear tendency (b) Low damping capacity
(c) High damping capacity (d) None of these

21. Head stock is situated at
(a) Left hand of the bed (b) Right hand of the bed
(c) Carriage (d) Lead screw

22. Outer surface of lathe spindle nose is generally threaded to mount
(a) Chuck (b) Face plate
(c) Both (a) and (b) (d) None of these

23. Tailstock is a
(a) Variable casting (b) Movable casting
(c) Fixed casting (d) All of these

24. Carriage is located between
(a) Headstock and saddle (b) Cross slide and saddle
(c) Tool post and apron (d) Head stock and tail stock

25. Carriage is used for ______ the tool
(a) Sharpening (b) Cutting
(c) Heating (d) Holding and feeding

26. Saddle is a ______ component.
(a) V shaped (b) U shaped
(c) I shaped (d) H shaped

27. Cross slide is mounted _______ of the saddle
(a) At bottom (b) On top
(c) On compound rest (d) On tool post

28. Compound rest is mounted on
(a) Cross slide (b) Tool post
(c) Apron (d) All of these

29. Apron is fastened to the
(a) Saddle (b) Head stock
(c) Tail stock (d) Legs

30. Split nut is used to move the _______ in thread cutting operation.
(a) Carriage (b) Lead stock
(c) Tail stock (d) Chuck

31. Removing the excess material from the outer surface to generate tapered surface is known as
(a) Taper cutting (b) Facing
(c) Taper turning (d) Turning

32. Compound rest method is used for turning
(a) Short tapers with large angle
(b) Small angle of taper on long workpiece

(c) Large taper with small angle
(d) All of above

33. Tail stock set over method is used for turning
(a) Short tapers with large angle
(b) Small angle of taper on long workpiece
(c) Large taper with small angle
(d) All of these

34. Operation required on lathe to produce flat surface at the end of workpiece is
(a) Boring (b) Reaming
(c) Parting (d) Facing

35. Process of enlarging a previously drilled hole is
(a) Drilling (b) Reaming
(c) Boring (d) Knurling

36. Operation of embossing a particular pattern on the surface of workpiece is known as
(a) Drilling (b) Reaming
(c) Boring (d) Knurling

37. In shaper, cutting action takes place in
(a) Forward stroke (b) Return stroke
(c) Idle stroke (d) All of these

38. Idle stroke in shaper is
(a) Return stroke (b) Forward stroke
(c) Cutting stroke (d) Both (a) and (b)

39. In shaper, feed is given to
(a) Tool (b) Workpiece
(c) Tool and workpiece both (d) None of these

40. Clapper box is provided on the shaper to prevent ______
(a) Clapping of tool (b) Damage of tool cutting edge
(c) Heating of tool (d) Cooling of tool

41. Clapper box raises the tool during
(a) Return stroke (b) Forward stroke
(c) Casting stroke (d) Both (a) and (b)

42. Tool head is fitted on the _____ of ram.
(a) Front end (b) Rear end
(c) Top (d) Bottom

43. In planer, cross rail supports the
(a) Tool head (b) Table
(c) Crown (d) Bed

44. In planer operations, feed is given to
(a) Tool (b) Workpiece
(c) Both (a) and (b) (d) None of these

45. Drilling machines are used
(a) To enlarge a hole (b) To produce a hole
(c) To finish hole to accurate size
(d) All of these

46. In drilling operation, hole is produced by feeding the drill
(a) Parallel to drill axis (b) Perpendicular to drill axis
(c) Both (a) and (b) (d) None of these

47. Which one is not a part of drilling machine
(a) Column (b) Base
(c) Clapper box (d) Table

48. Reaming operation is performed to **[UPTU Dec. 2010]**
(a) Enlarge a hole (b) Produce a hole
(c) Finish hole to accurate size (d) All of these

49. To correct the position of hole, operation required is
(a) Boring (b) Reaming
(c) Drilling (d) All of these

50. Counter sinking is the operation of
(a) Making cone shape enlargement
(b) Finishing a hole
(c) Squaring the surface around a hole
(d) Producing internal threads

51. Tapping is the operation of
(a) Making cone shape enlargement
(b) Finishing a hole
(c) Squaring the surface around a hole
(d) Producing internal threads

52. Spot facing is done for
(a) Making cone shape enlargement
(b) Finishing a hole
(c) Squaring the surface around a hole
(d) Producing internal threads

53. Up milling is also known as
(a) Conventional milling (b) Climb milling
(c) Unconventional milling (d) Both (a) and (b)

54. Down milling is also known as
(a) Conventional milling (b) Climb milling
(c) Unconventional milling (d) Both (a) and (c)

55. In up milling workpiece is fed in direction ______ cutter rotation.
(a) Opposite to (b) Same as of
(c) Perpendicular to (d) Parallel to

56. In down milling workpiece is fed in direction _____ cutter rotation.
(a) Opposite to (b) Same as of
(c) Perpendicular to (d) Parallel to

57. In up milling, chip thickness is _______ at beginning and ______ at the end.
(a) Maximum, minimum (b) Minimum, maximum
(c) Maximum, maximum (d) Minimum, minimum

58. In down milling, chip thickness is ________ at beginning and ______ at the end.
(a) Maximum, minimum (b) Minimum, maximum
(c) Maximum, minimum (d) Minimum, minimum

59. Slab or plain milling is done to produce horizontal flat surfaces
(a) Parallel to cutter axis (b) Perpendicular to cutter axis
(c) Both of above (d) None of above

60. Face milling is done to produce flat surfaces
(a) Parallel to cutter axis (b) Perpendicular to cutter axis
(c) Both of above (d) None of above

61. If more than two cutters are mounted on the same arbor than milling is
(a) Gang milling (b) End milling
(c) Straddle milling (d) Form milling

62. Irregular contours are produced by using form cutters in
(a) Gang milling (b) End milling
(c) Straddle milling (d) Form milling

63. Finishing is **[UPTU Dec. 2009]**
(a) Machining process (b) Welding process
(c) Joining process (d) Metal forming process

64. Material removal is a **[UPTU May 2010]**
(a) Machining process (b) Welding process
(c) Joining process (d) Metal forming process

C. Match the Following

1.

(A) Machine	(B) Features
(i) Lathe	(a) Dividing head
(ii) Drilling machine	(b) Rachet and paul mechanism
(iii) Shaper	(c) Hollow tapered spindle
(iv) Milling machine	(d) Face plate

2.

(A) Tool parts	(B) Machine
(i) Lead screw	(a) Lathe
(ii) Rocker arm	(b) Milling m/c
(iii) Universal indexing	(c) Shaper
(iv) Flute	(d) Drilling m/c

3.

(A)	(B)
(i) Permits angular movement of tool	(a) Cross slide
(ii) Holds the tool or tool holder with the tool	(b) Half nut
(iii) Permits transverse movement of the tool	(c) Tool post
(iv) Moves the whole carriage in conjuction with lead screw	(d) Compound rest

4.

(A)	(B)
(i) Drilling	(a) Finish the surface
(ii) Boring	(b) Produce circular hole

	(iii) Grinding		(c) Produce flat surface
	(iv) Facing		(d) Enlarge the drilled hole

5.

(A)	(B)
(i) Workpiece reciprocates past a stationary tool	(a) Gang milling
(ii) Tool reciprocates past a stationary workpiece	(b) Planer
(iii) More than two cutters	(c) Straddle milling
(iv) Two side milling cutters	(d) Shaper

ANSWERS

A. True or False

1. False	**2.** False	**3.** True	**4.** False	**5.** True
6. False	**7.** True	**8.** False	**9.** False	**10.** True

B. Multiple Choice Questions

1. (a)	**2.** (b)	**3.** (d)	**4.** (b)	**5.** (a)
6. (b)	**7.** (b)	**8.** (c)	**9.** (c)	**10.** (a)
11. (d)	**12.** (a)	**13.** (a)	**14.** (a)	**15.** (a)
16. (b)	**17.** (b)	**18.** (b)	**19.** (c)	**20.** (c)
21. (a)	**22.** (c)	**23.** (b)	**24.** (d)	**25.** (d)
26. (d)	**27.** (b)	**28.** (a)	**29.** (a)	**30.** (a)
31. (c)	**32.** (a)	**33.** (b)	**34.** (d)	**35.** (c)
36. (d)	**37.** (a)	**38.** (a)	**39.** (b)	**40.** (b)
41. (a)	**42.** (a)	**43.** (a)	**44.** (a)	**45.** (b)
46. (a)	**47.** (c)	**48.** (c)	**49.** (a)	**50.** (a)
51. (d)	**52.** (c)	**53.** (a)	**54.** (b)	**55.** (a)
56. (b)	**57.** (b)	**58.** (a)	**59.** (a)	**60.** (b)
61. (a)	**62.** (d)	**63.** (a)	**64.** (a)	

C. Match the Following

1. (i)—(d), (ii)—(c), (iii)—(b), (iv)—(a) **2.** (i)—(a), (ii)—(c), (iii)—(b), (iv)—(d)
3. (i)—(d), (ii)—(c), (iii)—(a), (iv)—(b) **4.** (i)—(b), (ii)—(d), (iii)—(a), (iv)—(c)
5. (i)—(b), (ii)—(d), (iii)—(a), (iv)—(c)

❑❑❑

CHAPTER 8

Welding

INSIDE THIS CHAPTER

8.1. INTRODUCTION

Process of joining two or more elements is known as fabrication. Joining may be temporary, permanent or semi-permanent. Joining by means of bolts, screws etc. are temporary fastening, by means of rivets is semi-permanent fastening and by welding is permanent joining.

Welding is the most extensively used method in industry and can be defined as, "The process in which two metals are joined permanently by heating upto fusion temperature with or without application of pressure is called welding. To weld the metal filler metal may or may not be used." The thermal energy required for welding is supplied by electrical or chemical means.

Advantages of Welding

1. A properly made weld is as strong as base metal.
2. Joint produced is permanent.
3. Equipments are portable and inexpensive.
4. Can join similar and dissimilar metals.
5. it can be mechanized.

8.2. CLASSIFICATION OF WELDING [UPTU Dec. 2008, May 2009]

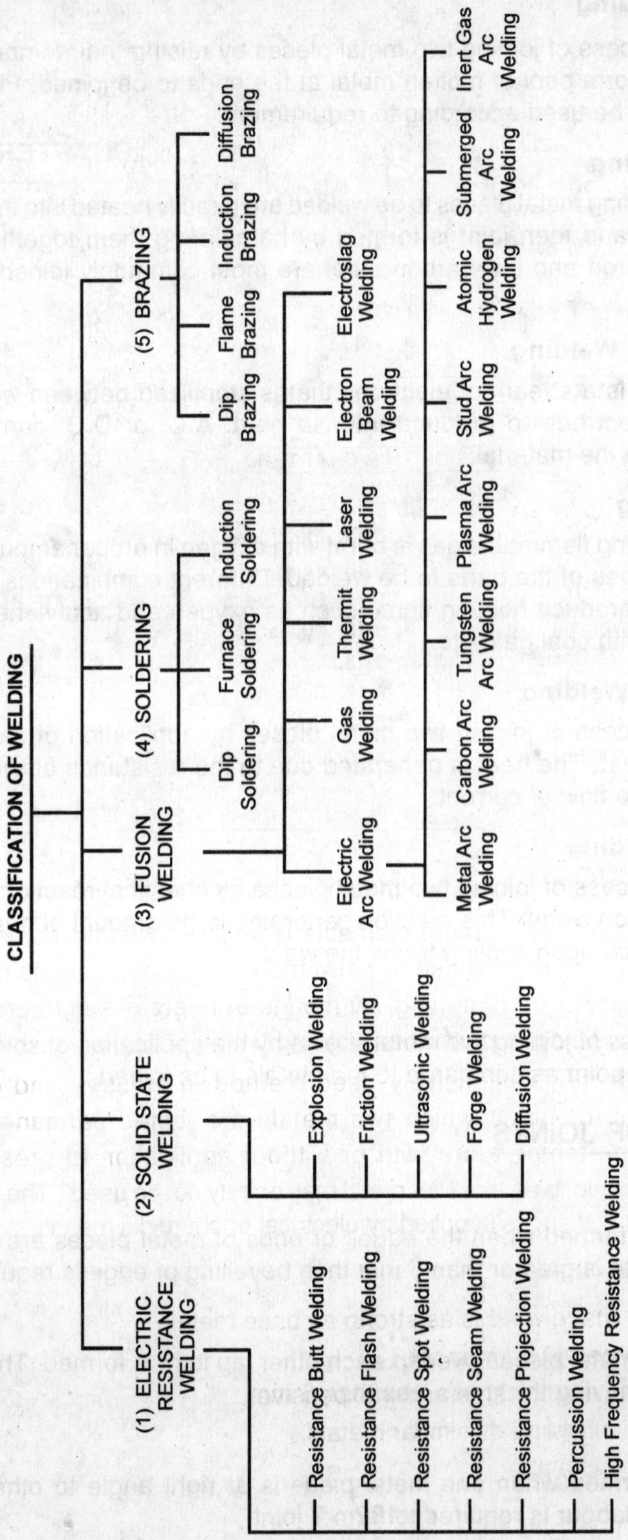

Fig. 8.1. Classification of Welding.

Main types of welding process are as follows :

1. Fusion Welding

It is the process of joining two metal pieces by raising their temperature to fusion point to form pool of molten metal at the ends to be joined. Filler metal may or may not be used according to requirement.

2. Forge Welding

In forge welding metal pieces to be welded are directly heated into the furnace to plastic state and then joint is formed by hammering them together on the anvil. Wrought iron and low carbon steel are most commonly joined by forge welding.

3. Electric Arc Welding

Electric arc is a stream of electrons that is stabilized between work piece and welding electrode to produce intense heat. A.C. or D.C. can be used depending upon the material.

4. Gas Welding

In this welding flammable gas is burnt with oxygen in proper amount to fuse together the edges of the parts to be welded. Different combinations are used with oxygen to produce heating flame such as oxygen and acetylene, oxygen and hydrogen with coal gas, etc.

5. Resistance Welding

It is the process of joining two metal pieces by application of mechanical pressure and heat. The heat is generated due to the resistance offered by the two pieces to the flow of current.

6. Thermit Welding

It is the process of joining two metal pieces by chemical reaction between aluminum and iron oxide. This reaction generates large amount of heat to form molten pool which upon cooling forms the weld.

7. Soldering

It is a process of joining two metal sheets by the application of solder which has low melting point as compared to the metals to be joined.

8.3. TYPES OF JOINTS

1. Butt Joint

Butt joint is formed when the edges or ends of metal pieces are joined. If thickness of plate is greater than 5 mm then bevelling of edge is required.

2. Lap Joint

When both metal pieces overlap each other lap joint is formed. This joint is used for plates having thickness less than 3 mm.

3. T Joint

T joint is formed when one metal piece is at right angle to other at mid section. Skilled labour is required to form T joint.

4. Corner Joint

Corner joint is formed when one metal piece is at right angle to other at corner. These joints are less costly due to less use of filler metal.

5. Edge Joint

This joint is used to join the edge of two pieces in which a part of the surface remains parallel to each other. These joints are economical due to less requirement of edge preparation and less use of filler metal.

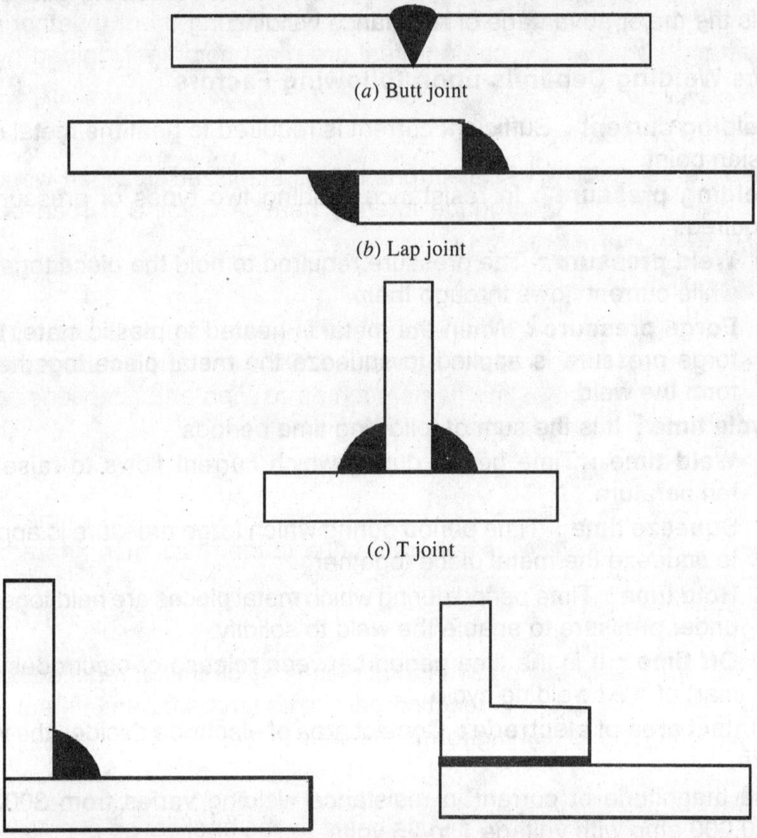

(*a*) Butt joint

(*b*) Lap joint

(*c*) T joint

(*d*) Corner joint (*e*) Edge joint

Fig. 8.2. Types of Joints.

8.4. WELDABILITY

It indicates the ease with which two similar or dissimilar metals can be welded with desired toughness, strength and without any welding defect and residual stresses. Various factors which affect the weldability are :

- Welding process
- Surface condition.
- Properties of metal to be welded like melting point, thermal conductivity, thermal expansion coefficient, etc.
- Filler metal to be used.
- Composition of the metal.

8.5. RESISTANCE WELDING [UPTU MAY 2009]

It is the process of joining two metal pieces by application of mechanical pressure and heat. The heat is generated due to the resistance offered by the two pieces to the flow of current. In resistance welding two metal pieces to be joined are held together and strong electric current is passed through it. Due to flow of current across the resistance, the temperature at the junction reaches the fusion point. At this moment slight mechanical pressure is applied to complete the weld. In this welding consumable electrode, flux and shielding gas is not used that is the major advantage of resistance welding.

Resistance Welding Depends upon following Factors :

1. **Welding current :** Sufficient current is required to heat the metal upto fusion point.
2. **Welding pressure :** In resistance welding two types of pressure is required.
   (a) **Weld pressure :** The pressure required to hold the piece together while current flows through them.
   (b) **Forge pressure :** When the metal is heated to plastic state, then forge pressure is applied to squeeze the metal piece together to form the weld.
3. **Cycle time :** It is the sum of following time periods
   (a) **Weld time :** Time period during which current flows to raise the temperature.
   (b) **Squeeze time :** Time period during which forge pressure is applied to squeeze the metal piece together.
   (c) **Hold time :** Time period during which metal pieces are held together under pressure to enable the weld to solidify.
   (d) **Off time :** It is the time period between release of electrodes and start of next welding cycle.
4. **Contact area of electrode :** Contact area of electrode decides the weld size.

   The magnitude of current in resistance welding varies from 3000 to 100,000 amp with voltage 1 to 25 volts.

   Heat generated in resistance welding is given by general expression

$$H = I^2Rt$$

where,

H = Heat generated, in joules
I = Current, in amperes
R = Resistance, in ohms
t = Time of current flow, in seconds

8.6. CLASSIFICATION OF RESISTANCE WELDING

Resistance welding can be further sub-divided as

1. Spot Welding
2. Butt Welding

3. Flash Welding
4. Seam Welding
5. Projection Welding
6. Percussion Welding

8.6.1. Resistance Spot Welding

Spot welding is a resistance welding process in which overlapping metallic plates are held between two copper electrodes, which concentrate welding

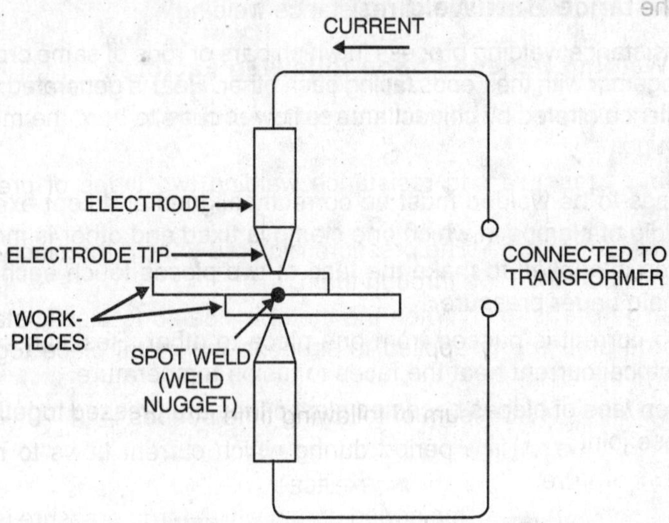

Fig. 8.3. Resistance Spot Welding.

current to melt the interface over a spot and apply pressure to complete the weld. In this welding accurate control and timing of electrical current and pressure are essential.

Procedure

- Clean the job and electrode tip.
- Press the metal sheets between the electrodes.
- Flow high current at low voltage through the metal pieces by means of transformer. Temperature at the contact surfaces rises to fusion point due to the resistance offered by the sheet metal to the flow of current and weld is completed under the contact pressure of electrodes.

Features

- Current is of the order of 3000 to 100,000 ampere.
- Temperature of weld zone is 815°C to 930°C.
- Electrodes are made up of copper base alloys and refractory metal alloys.
- Weld nugget is generally 6-10 mm in diameter.

Advantages

- High speed of welding.
- Less skilled worker is required.

- No edge preparation is needed.
- Warping or distortion is eliminated.

Applications

- For welding lap joint having plate thickness upto 12.5 mm thickness.
- Used for ferrous and non-ferrous alloys.
- Used for welding of automobile bodies, refrigerator and other metal stamping assemblies.

8.6.2. Resistance Butt Welding

It is a resistance welding process in which bars or rods of same cross-section are joined together with their ends facing each other. Heat is generated at the ends due to resistance offered by contact area to flow of current.

Procedure

- Pieces to be welded must be correctly aligned and kept exactly in the middle of clamps in which one clamp is fixed and other is moving.
- Force is applied to make the face of two pieces touch each other and remain under pressure.
- High current is passed from one piece to other. Resistance to flow of electrical current heat the faces to fusion temperature.
- When face of pieces becomes plastic they are pressed together to form dense joint.

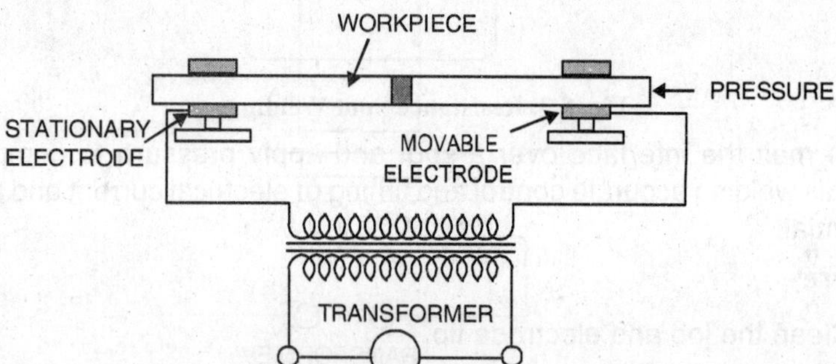

Fig. 8.4. Resistance Butt Welding.

Features

- Clamps are made of Cu alloy.
- Step down transformer is used to provide the electrical current for heating.
- Parent metal is used as electrode.

Applications

- Welding of ferrous and non-ferrous rods.
- Welding of longitudinal butt joints in tubes and pipes.
- Welding of transverse butt joints in steel rings.

8.6.3. Resistance Flash Welding

It is used to make butt joints by welding metal pieces of dissimilar metal and can be used for all metal that can be forged easily. It is generally used in place of butt welding for articles having thin cross-section. In flash welding current is passed while the workpiece are still separated to establish flashing.

Procedure

- Both metal piece are properly aligned and fixed in the middle of clamps.
- Current is switched on, flashing is established and ends of pieces to be welded are brought closer.
- As the flashing continues, the temperature of ends of metal piece increases and attain welding temperature.
- At this moment mechanical pressure is applied which forces the molten metal and slag to be expelled out in form of sparks enabling a good solid joint.

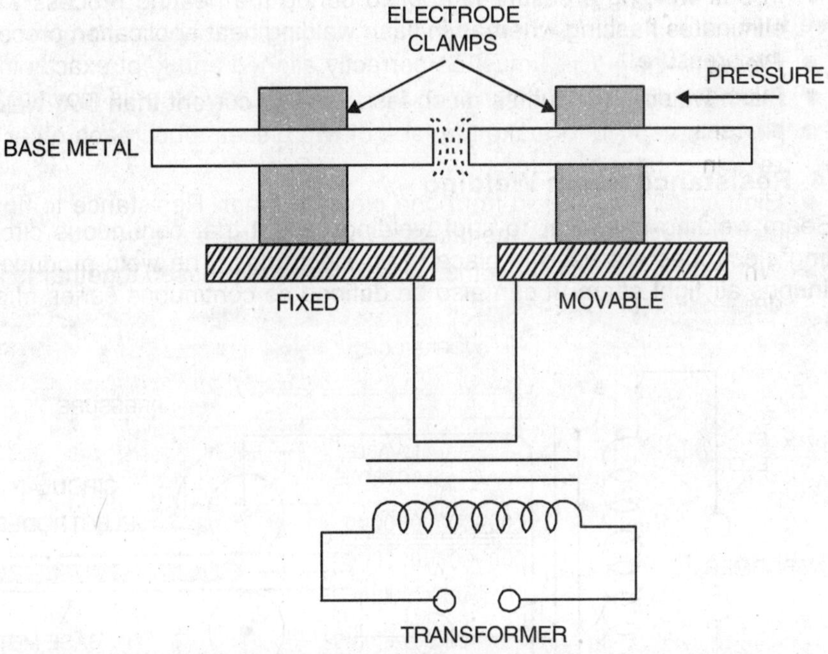

Fig. 8.5. Resistance Flash Welding.

Advantages

- Require less current as compared to butt welding.
- It is much faster than butt welding.
- It is much stronger than butt welding.
- Ends need not be squared, which is the primary requirement of butt welding.
- Quality of weld is good due to expelling of the impurities and contaminants during the operation.

Disadvantages

- Highly skilled labour is required.
- Additional stock has to be provided to compensate loss of metal during flashing.

- During flashing, particles of molten metal are thrown out, which may enter into the slideways and insulation.
- Risk of fire hazard.

Applications

- Welding of metal sheets, tubes, bars, rods, etc.
- Welding of tubular and solid structure assemblies.
- Welding of tool steel drill, tap and reamer bodies.
- In automobile and aircraft industry, household appliances, etc.

Difference between Flash and Butt Welding

- In butt welding no flashing takes place between the surfaces.
- In butt welding pressure is applied during the heating process which eliminates flashing whereas in flash welding heat application precedes the pressure.
- Flash welding consumes much less welding current than butt welding process.

8.6.4. Resistance Seam Welding

Seam welding is similar to spot welding except that continuous circular rotating electrodes are used in place of tip electrodes. The weld produced is continuous air tight seam. It can also be defined as continuous series of spot welds.

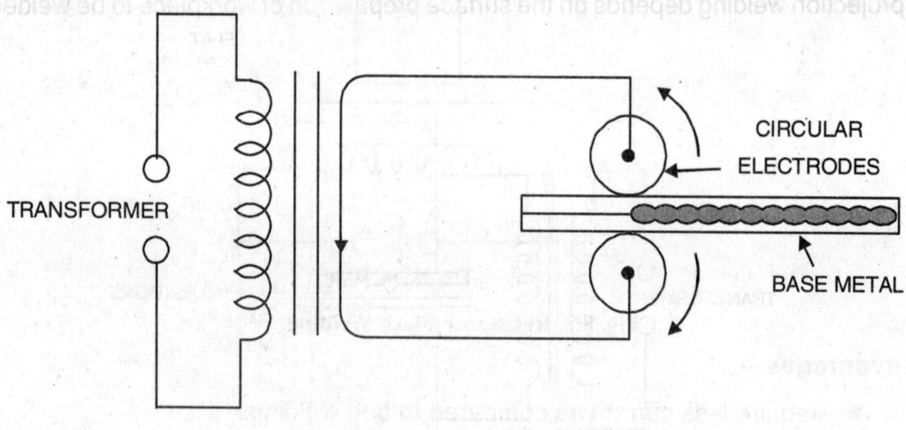

Fig. 8.6. Resistance Seam welding.

Procedure

- Workpiece are cleaned, overlapped and placed between two circular electrodes that clamp the workpiece together.
- Current is switched on and metal pieces are pushed together to travel between rotating electrodes.
- The metal pieces, under the constant pressure of rotating electrodes get heated to plastic state and welded continuously as it passes between them.

Advantages

- It can produce gas tight or liquid tight or steam tight joints.
- A single seam weld or several parallel seam welds may be produced simultaneously.
- It is quicker operation than spot welding and gives a stronger joint.

Disadvantages

- Weld should proceed along a straight line.
- Can not be applied to sharp corners.
- There should be no obstruction in path of electrode.
- Work surface should be clean to produce defect free welding.

Applications

- In welding of metals other than copper and copper alloys.
- To made Girths welds in round, square or rectangular form.
- In producing pressure tight weld joints on gasoline tank, oil drums, transfer tank, etc.

8.6.5. Resistance Projection Welding

It is developed form of spot welding in which the spots, at which welding takes place are previously located by providing projections at desired location on the surface of one of the work piece. Electrodes, instead of having tip as in spot welding are flat and relatively larger in surface area. The success of projection welding depends on the surface preparation of workpiece to be welded.

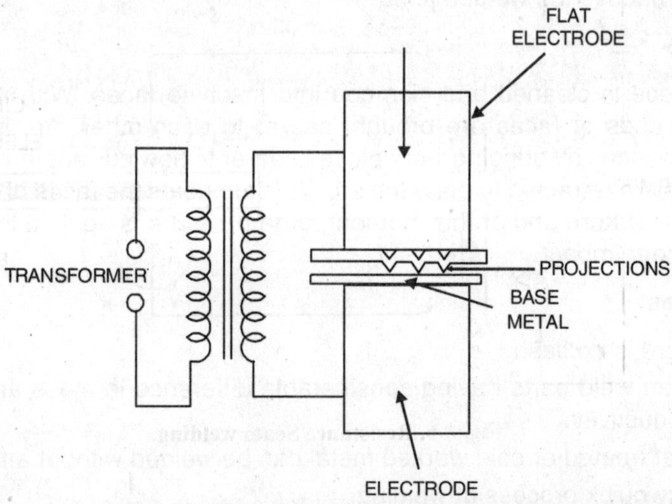

Fig. 8.7. Resistance Projection Welding.

Procedure

- Projection in upper piece is held in contact with lower piece under electrode pressure.
- As the current flows, the region around the projection gets heated up and reaches to plastic state.
- At this moment slight pressure is applied by electrode to complete the weld.

Advantages

- It can weld thick metals.
- It enables the production of several spot welds simultaneously.
- Closer welds can be obtained by this process.
- It lowers the amount of current and pressure needed to form a good bond between two surfaces.
- It possess longer life than spot welding.

Disadvantages

- Metals which are not strong enough to support projections and withstand electrode force are not be welded satisfactorily.
- All projection must be of same height to produce a good weld.
- It is economic only when the production rate and volumes are large.

Applications

- Smaller fasteners can be welded to large components.
- In production of light steel frames.
- In automobile industry.

8.6.6. Percussion Welding

In this process intense and highly concentrated heat is obtained by an electric arc produced by rapid discharge of stored electrostatic energy. The intense heat brings the ends to a molten state and at this time percussion of one part against the other produces the welded joint.

Procedure

Workpiece is cleaned and clamped into machine faces. With the help of light force, ends or faces are brought nearer to each other. Arc is strucked between the faces by bringing work pieces nearer to flow current through them. Workpieces are retracted to draw the arc. This arc heats the faces of workpiece to weld temperature and at this moment forge pressure is applied to complete the weld under impact.

Advantages

- There is no flash.
- It can weld parts having considerable difference in mass and thermal conductivity.
- Heat treated or cold worked metal can be welded without annealing.
- Very quick process of welding.

Disadvantages

- Process is limited to welding of butt joints only.
- Accurate holding fixture and sensitive timing devices are required.

Applications

- Used for welding stellite tips to tools.
- Used to join similar and dissimilar metals.

- Used in telephone industry for connecting leaded components to terminals.
- Used in welding of filaments.

8.7. ELECTRIC ARC WELDING

In electric arc welding metal pieces to be joined are heated to melting point by creating an electric arc between them to form a pool of molten metal and then allowing it to solidify to form the welded joint. Sometimes additional material is added to form the weld by melting the wire known as filler metal.

Electric arc welding is of two types :

1. Carbon Arc Welding
2. Metal Arc Welding

8.7.1. Carbon Arc Welding

This welding is suitable for joining steel sheets and repairing of steel castings. In this welding only D.C. with straight polarity is used which restrict electrode disintegration and amounts of carbon going into the weld work. High voltage D.C. generator or transformer with rectifier, are used to supply D.C upto 600 ampere. Arc is strucked between job and carbon electrode and allowed to impinge on the surface to be welded till molten pool forms. Filler metal and flux may or may not be used.

8.7.2. Metal Arc Welding

In metal arc welding arc is maintained between electrode and workpiece which forms the two terminals. In this welding, electrodes used may be bare or coated. Bare electrode has same composition as that of parent metal whereas coated electrodes have coating of some material or flux that prevents the oxidation of surface.

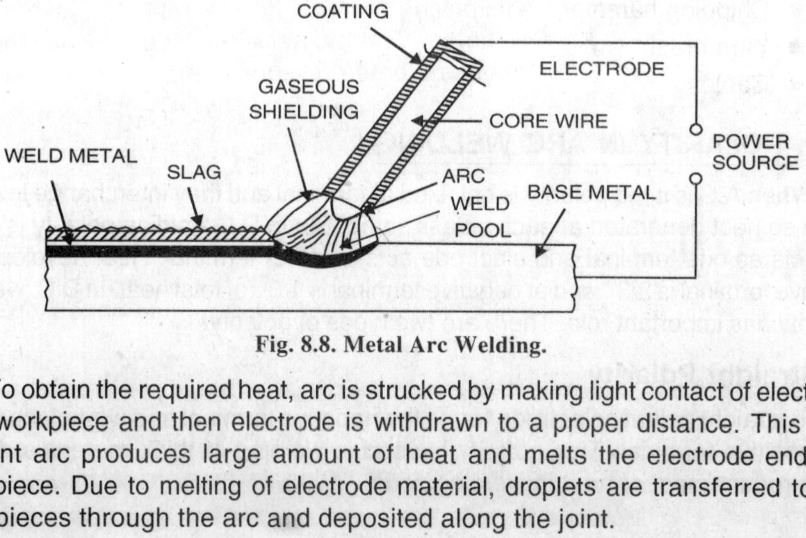

Fig. 8.8. Metal Arc Welding.

To obtain the required heat, arc is strucked by making light contact of electrode with workpiece and then electrode is withdrawn to a proper distance. This high current arc produces large amount of heat and melts the electrode end and workpiece. Due to melting of electrode material, droplets are transferred to the workpieces through the arc and deposited along the joint.

Metal electrode is coated with flux which produces a gas shield around the arc to protect it from atmosphere. Metal flux also produces slag to prevent atmosphere contamination of molten weld metal.

Metal arc welding is suitable for welding of plates having thickness 3 mm to 20 mm. It is not suitable for small thickness (less than 3 mm) because of lack of rigidity and for large thickness (more than 20 mm) because of taking long time for filling up the joint groove.

Arc crater : Due to the blast of arc, molten metal is forced out of the pool and forms a small depression in the parent metal. This depression caused due to the penetration of arc is known **arc crater.**

8.8. ARC BLOW

In D.C welding magnetic fields are setup around the electrode and workpiece due to flow of current. When these magnetic field are unbalanced they fluctuate the arc and bend it away from its path. The deflection of arc from its path is called arc blow.

8.9. ARC WELDING EQUIPMENTS

In arc welding both A.C and D.C power sources are used. In D.C. welding D.C generator driven by electric motor or engine is used to generate D.C. In A.C. arc welding a step down transformer is used which converts 400-440 volt supply to 80-100 volt suitable for welding.

Other welding equipments requried in arc welding are :

- Electrode holder
- Welding leads
- Safety goggles
- Ground connection
- Cable lugs
- Hand shields
- Helmet, gloves
- Chipping hammer
- Wire brush
- Cables

8.10. POLARITY IN ARC WELDING

When A.C is used polarity is not fixed at terminal and they interchange in every cycle so heat generated at each pole is same. But in D.C welding polarity is fixed, job acts as one terminal and electrode acts as other terminal. Heat developed at positive terminal is $2/3^{rd}$ and at negative terminal is $1/3^{rd}$ of total heat. In D.C. welding polarity has important role. There are two types of polarity.

1. Straight Polarity

In straight polarity electrode forms the negative terminal and workpiece forms the positive terminal. This polarity is used in welding of thick materials due to large requirement of heat in welding zone.

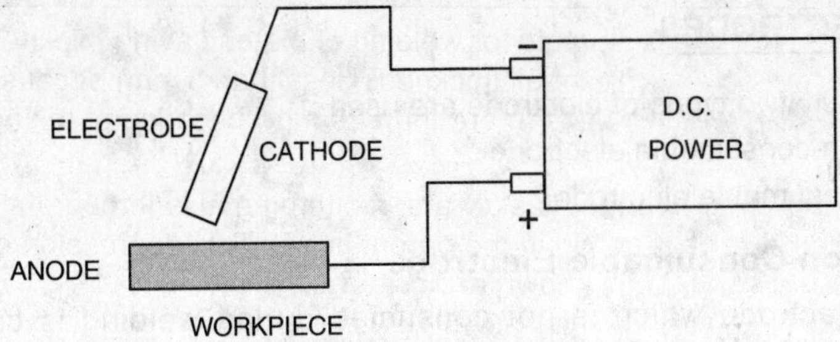

Fig. 8.9. Straight Polarity.

2. Reverse Polarity

In reverse polarity electrode forms the positive terminal and workpiece forms the negative terminal. This polarity is used in welding of thin materials due to less requirement of heat in welding zone.

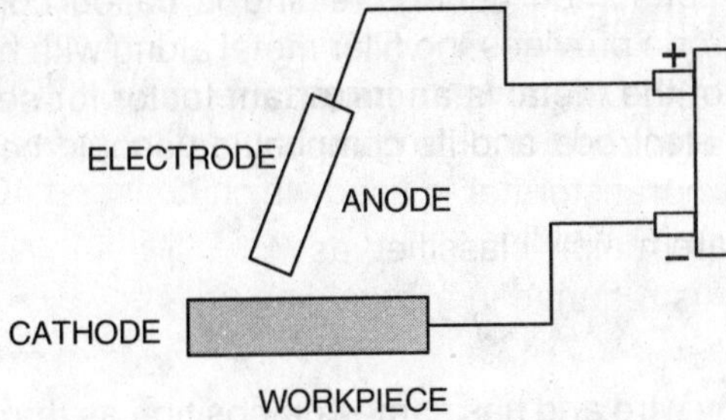

Fig. 8.10. Reverse polarity.

8.11. COMPARISON BETWEEN A.C. AND D.C. WELDING

S.No	A.C. Welding	D.C. Welding
1.	Welding with A.C. is cheap and easy.	Welding with D.C. is costly and difficult.
2.	Straight polarity and reversed polarity are not possible.	Straight polarity and reversed polarity both are possible with D.C.
3.	Used for welding of ferrous metal.	Welding of both ferrous and nonferrous metal can be done.
4.	Problem of arc blow doesn't exist.	Problem of arc blow exist
5.	Striking of arc is difficult.	Striking of arc is relatively easy.
6.	Equal amount of heat is generated at both poles.	More heat is generated at positive pole and less heat is generated at negative pole
7.	It is difficult to maintain and start the arc as voltage becomes zero twice in a cycle.	Voltage remains constant throughout the cycle.
8.	Not preferred for welding of sheet metal	It is preferred for welding of sheet metal.

8.12. ELECTRODES

Generally two types of electrode are used :

1. Non-consumable electrode
2. Consumable electrode

8.12.1. Non Consumable Electrode

The electrode which is not consumed during welding is called non consumable electrode. Filler rods are used separately with non consumable electrode to fill the gap along the joint. These electrodes are made of carbon, graphite or tungsten. Tungsten electrodes are used for A.C. and D.C. welding whereas graphite and carbon are used only for D.C. welding.

8.12.2. Consumable Electrode

The electrode which is consumed during welding is called consumable electrode. Consumable electrode provides the filler metal along with heat to the joint. Chemical composition of the metal is an important factor for selection of the material for consumable electrode and its composition should be same as that of material to be welded.

Consumable electrodes are mainly classified as

1. Bare Electrodes

It consists of metal or alloy wire and has same composition as that of parent metal. During welding with bare electrode nonmetallic constituents are trapped in weld metal and decreases the strength and ductility of metal so it is rarely used.

2. Coated Electrodes

When core of bare metallic wire is provided with a coating or covering on the outside surface then the electrode is known as coated electrode. The weight of coating is 1.5 to 30% of weight of electrode rod.

8.13. FLUX

Flux are used to dissolve oxides and impurities for successful welding. It should have lower specific gravity and melting point than base and filler metal.

Electrodes are covered with flux to dissolve the oxides present in the surface of the metal. Flux coated electrodes acts as deoxidizer, provide protective atmosphere, improve stability of arc and reduces weld metal spatter. Borax and sodium carbonate are good fluxes for ferrous materials.

8.14. GAS WELDING

It is a fusion welding process that joins the metal by using heat of combustion of strong gas flame. The intense heat raises the temperature of ends, melts and fuses together the edges of the parts to be welded. Filler metal may be added to flowing molten metal to fill the cavity. Different combinations are used to produce different heating flames e.g. oxygen and acetylene, oxygen and hydrogen with coal gas, etc.

8.14.1. Oxy Acetylene Welding

It is most commonly used for welding of almost all metals and alloys. Acetylene is mostly used with oxygen because it produces high temperature and form inert gas envelope to prevent oxidation of molten metal. When oxygen is mixed with acetylene in correct proportions in the welding torch and ignited, the flame resulting at the tip of torch is sufficiently hot to melt and join the parent metal. The oxyacetylene flame attains approximately a temperature of 3200°C.

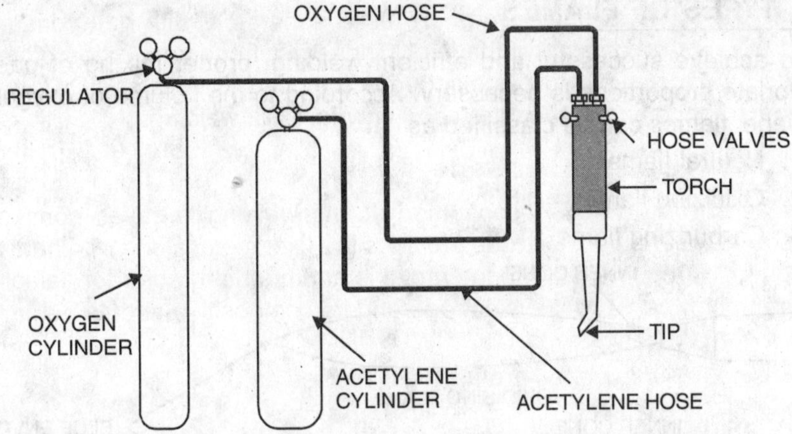

Fig. 8.11. Oxy Acetylene Welding.

8.14.2. Equipments

1. Oxygen Gas Cylinder

It is painted with black colour and is filled with oxygen at a pressure of 125 Kgs to 140 kgs per square centimeter. A removable steel cap is screwed on the cylinder when cylinder is not in use.

2. Acetylene Gas Cylinder

It is painted with maroon colour. Cylinder is filled with spongy material saturated with chemical solvent known as acetone. Acetone has the ability to dissolve 25 times its own volume of acetylene per atmosphere of pressure.

3. Blowpipe or Welding Torch

Welding torch consists of two different passages, one for oxygen and other for acetylene. It consists of welding tip at the end where flame is generated. Pressure regulators on welding torch are used to vary the supply of gases to generate different types of flames.

4. Pressure Regulator

The purpose of pressure regulator is to produce a steady flow of gas under varying pressure. It is also used to adjust the pressure of gas according to requirement.

5. Hose Pipe

Hose pipe is used to carry gases from cylinder to welding torch. Red colour pipe is used to carry acetylene and blue colour pipe is used for oxygen. It should be strong enough to resist gas pressure.

6. Stop Valve

It is used to stop or start the supply of gas to welding torch.

7. Pressure Gauge

It indicates the instantaneous pressure and remaining quantity of gas in respective cylinder at any instant. It is mounted on each cylinder.

8.15. TYPES OF FLAMES

To achieve successful and efficient welding, proper mixing of gases in appropriate proportions is necessary. According to the volume of oxygen and acetylene, flames can be classified as

1. Neutral flame
2. Oxidizing flame
3. Carburizing flame

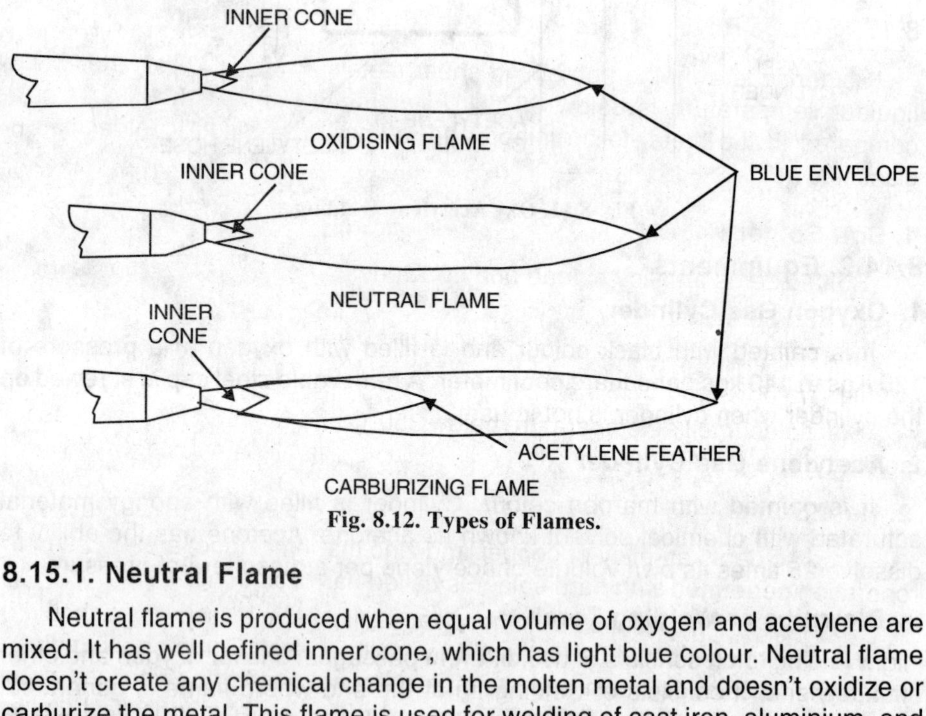

Fig. 8.12. Types of Flames.

8.15.1. Neutral Flame

Neutral flame is produced when equal volume of oxygen and acetylene are mixed. It has well defined inner cone, which has light blue colour. Neutral flame doesn't create any chemical change in the molten metal and doesn't oxidize or carburize the metal. This flame is used for welding of cast iron, aluminium, and stainless steel. The temperature of neutral flame is approximately 3260°C.

8.15.2. Oxidizing Flame

Oxidizing flame is obtained by increasing the supply of oxygen in mixture. It has small white cone that is shorter and more pointed than that of neutral flame. It creates high temperature of 3482°C. It is normally not used except in case of brass.

8.15.3. Carburizing Flame

It is produced when supply of acetylene is higher than supply of oxygen and can create an approximate temperature of 3038°C. It can be identified by acetylene feather. Its outer envelope is longer than that of neutral flame. It is used for welding of lead.

In these flames, temperature is maximum at the top of inner cone. Feather or brush exist next to the cone and occurs in carburizing flame only.

8.16. COMBUSTION OF GAS

Combustion of gas takes place in two stages:

Stage 1: In this stage, oxygen combines with acetylene to form carbon monoxide. Hydrogen is liberated in this reaction.

$$C_2H_2 + O_2 \longrightarrow 2CO + H_2$$

Stage 2: Carbon monoxide combines with oxygen of surrounding air to form carbon dioxide, and hydrogen burns with oxygen to form water vapour.

$$2CO + H_2 + 3O \longrightarrow 2CO_2 + H_2O$$

8.17. SOLDERING

Soldering is a method of joining sheet metals by using filler metal whose liquidus temperature is below **427°C**. Filler metal has low melting point as compared to the metal to be joined and is known as solder. Solder can be classified as :

1. Soft Solder

Soft solder is an alloy of lead and tin. Generally used compositions are :

(i) Tin 50%, lead 50%,
(ii) Tin 67% lead 33%
(iii) Tin 33% lead 67%

Soft solder melt at temperature below 350°C and operation is known as soft soldering.

2. Hard Solder

Hard solder is an alloy of copper and zinc that melts above 600°C and the operation performed with hard solder is called hard soldering.

To prevent the oxidation of joint surfaces, fluxes are used. Flux should be light in weight so it could be displaced by molten metal. Common examples of flux are zinc chloride, rosin and rosin plus alcohol based flux, mixture of zinc chloride and ammonium chloride. Rosin and rosin plus alcohol based flux are used in electrical soldering work.

Procedure of Soldering

1. Clean the surfaces to be joined.
2. Place the parts facing each other and apply the flux.
3. Heat the soldering iron either electrically or in a furnace.
4. Bring the bit of soldering iron in contact with solder to melt and spread over the surface of bit.
5. This molten solder is applied on the surface with soldering iron to join them.

Advantages of Soldering

1. Soldering is used for joining wires and small parts.
2. Soldering is used in electrical components like printed circuit boards, television, transistor, etc.
3. It is used for coating and joining of metal, body soldering for filling in seams and dents in automobile bodies.

8.18. BRAZING

Brazing is a process in which metal pieces are joined by means of filler metal whose liquidus temperature is above **427**°C but below solidus temperature of base metal. This filler metal is called spelter and is a mixture of tin, copper and zinc. Brazing can join dissimilar metal such as cast iron or wrought iron metals, nonmetals to metals, etc.

Procedure

1. First of all clean the joints and apply the flux to avoid oxidation at high temperature. Borax is generally used as flux.
2. Clamp the part with proper clearance for filler metal.
3. Heat the parts above the melting point of spelter. Spelter together with flux melts and flows by capillary action along the contacting surface.
4. Allow the joint to cool and solidify to form the joint.

The clearance between two parts to be joined has important role. If clearance is large, the capillary action will not be able to draw the filler metal into joint. If clearance is small, then small amount of filler will provide ineffective strength.

Applications

1. Brazing is used to joint cast parts to wrought parts, dissimilar metals, nonmetals to metals.
2. Brazing is used to join carbide tips with mild steel shanks.
3. Brazing is also used for fastening of pipe fittings and electrical repair work.

Advantages

1. Brazing is used to fabricate large assemblies in stress free condition.
2. Brazing is used to join cast material to wrought metals.
3. Brazing can joint thin walled tubes and light gauge sheet metal assemblies which is not possible by welding.
4. Brazing is used to join large variety of dissimilar metals.

Disadvantages

1. Very large assemblies formed by brazing is costlier as compared to welding.
2. Brazed joints have poor strength than welding.
3. If flux residues are not properly removed corrosion may occur.
4. Brazing flux may produce toxic fumes.

8.19. DIFFERENCE BETWEEN WELDING AND RIVETING

Welding	Riveting
1. Welded structure possesses better finish and appearance as comapred to riveted structure.	1. Riveted structure possess poor finish and appearance as compared to welded structure.
2. Welding can be done at any point on a structure, requires less clearance.	2. Riveting require enough clerance.
3. It is an inexpensive operation.	3. It is an expensive operation.
4. It is a fast operation.	4. It is a slow operation.
5. Welding operation forms strong joint.	5. It forms weak joint as compared to welded joint.
6. It creates less noise as compared to riveting.	6. It is a noisy operation.
7. Complicated structures can be obtained by welding.	7. It is difficult to design complicated structure by riveting.
8. Welded structure can be easily modified.	8. It is very difficult to modify riveted structure.
9. Welded pressure vessel are more pressure tight as compared to riveted ones.	9. Riveted pressure vessel are less pressure tight.
10. Welded structure are light	10. Riveted structure are heavy as comapred to welded structure.

SUMMARY

- The process in which two metals are joined permanently by heating upto fusion temperature with or without application of pressure is called welding. To weld the metal filler metal may or may not be used.
- Weldability indicates the ease with which two similar or dissimilar metals can be welded with desired toughness, strength and without any welding defect and residual stresses.
- Resistance welding is the process of joining two metal pieces by application of mechanical pressure and heat generated due to the resistance offered by the two pieces to the flow of current.
- Resistance welding can be classified as Spot, Butt, Flash, Seam, Projection and Percussion Welding.
- In electric arc welding metal pieces to be joined are heated to melting point by creating an electric arc between them, and allowing them to solidify to form the welded joint.
- In straight polarity electrode forms the negative terminal and workpiece forms the positive terminal. In reverse polarity electrode forms the positive terminal and workpiece forms the negative terminal.
- Bare electrode has same composition as that of parent metal whereas coated electrodes are provided with a coating on bare metallic wire.

- Gas welding is a fusion welding process that joins the metal by using heat of combustion of strong gas flame.
- In neutral flame, equal volume of oxygen and acetylene are mixed. In oxidizing flame supply of oxygen is higher than acetylene whereas in carburizing flame supply of acetylene is higher than supply of oxygen.
- Soldering is a method of joining sheet metals by using filler metal whose liquidus temperature is below 427°C.
- Brazing is a process in which metal pieces are joined by means of filler metal whose liquidus temperature is above 427°C but below solidus temperature of base metal.

SHORT QUESTIONS WITH ANSWERS

Q. 1. What is the difference between joints formed by riveting and welding?

Ans. Riveting forms semi-permanent joint whereas welding forms permanent joint.

Q. 2. What is the difference between resistance butt welding and flash welding?

Ans. (1) No flash is formed in butt welding.

(2) Flash welding consumes less welding current than butt welding.

(3) In flash welding ends need not be squared which is primary requirement of butt welding.

Q. 3. What is the use of straight polarity and reverse polarity?

Ans. Straight polarity is used in welding of thick material and reverse polarity is used in welding of thin material.

Q. 4. List functions of flux.

Ans. Flux acts as deoxidiser, provide protective atmosphere, improve stability of arc and reduces weld metal spatter.

Q. 5. In which form acetylene is filled in acetylene cylinders?

Ans. Acetylene cylinder is filled with spongy material saturated with chemical solvent known as acetone.

Q. 6. Which materials are welded by neutral flame? **[UPTU May 2012]**

Ans. Cast iron, aluminium and stainless steel are welded by neutral flame.

Q. 7. What are the factors on which resistance welding depends?

Ans. (1) Welding current

(2) Cycle time

(3) Welding pressure

(4) Contact area of electrode.

Q. 8. What equipments are used in electric welding?

Ans. (1) Power source

(2) Electrode holder

(3) Welding leads

(4) Safety goggles

(5) Cable lugs

(6) Hand shields

Q. 9. Which material is used to made non-consumable electrode?

Ans. (1) Carbon, (2) Graphite and (3) Tungsten

Q. 10. What are the applications of resistance flash welding?

Ans. Flash welding is used in

(1) Automobile and aircraft industry.

(2) Welding of tool steel drill, tap and reamer bodies.

(3) Welding of tubular and solid structure assemblies

Q. 11. What is HAZ? **[UPTU Dec. 2011]**

Ans. The portion of base metal that has not been melted but whose microstructure or mechanical properties has been changed by heat of welding, is called heat affected zone (HAZ).

Q. 12. Write two difference between soldering and brazing. **[UPTU Dec. 2011]**

Ans. (1) In solding melting point of filler metal is below 427°C but in brazing melting point is above 427°C.

(2) In solding tin and lead alloys are used as filler metal but in brazing copper and silver base alloys are used.

REVIEW QUESTIONS

1. Define welding. Classify different welding processes. **[UPTU May 2010]**

2. Explain the principle of resistance welding. What factors are required to complete the resistance welding? **[UPTU May 2010]**

3. Classify the different resistance welding processes and explain them with neat sketches.

4. Explain butt welding and seam welding with neat diagrams.

5. What is principle of gas welding? **[UPTU Dec. 2010]**

6. What are the equipments used in oxyacetylene welding? Explain in detail.

7. Explain different types of oxyacetylene flames. **[UPTU Dec. 2010]**

8. What do you mean by electric arc welding? How is it classified? **[UPTU Dec. 2010]**

9. Explain principle of metal arc welding in detail with neat sketches.

10. Define

(a) Arc crater

(b) Arc blow

(c) Arc welding electrodes

11. List the various equipments used in arc welding and explain its advantages and disadvantages. **[UPTU Dec. 2010]**

12. Define straight polarity and reverse polarity.

13. Compare the use of A.C and D.C in arc welding.

14. What do you mean by soldering and brazing?

15. Distinguish beween welding, brazing and soldering processes. Write about the importance of fluxes being used in welding. **[UPTU Dec. 2009]**

16. How is an arc obtained in arc welding. What are different power sources used in welding. What are advantages and limitations of each. **[UPTU May 2012]**

SELF TEST QUESTIONS

A. State Whether True or False

1. Welding is a temporary fastening.
2. High melting point metal is used to join the sheet metal.
3. The time period during which metal piece are held together under pressure to enable the weld to solidify in resistance welding is known as hold time.
4. Resistance welding is used to from lap joint.
5. Flash welding consumes more current than butt welding.
6. The composition of bare electrode is not same as that of parent metal.
7. In A.C arc welding step down transformer is used
8. Heat developed at positive terminal is one third of total heat.
9. In D.C welding equal amount of heat is generated at both poles.
10. Oxygen gas cylinder is painted with black colour.
11. The filler metal used in solder has high melting point as compared to parent metal to be joined.
12. In A.C welding, problem of arc blow exist.

B. Fill In The Blanks

1. Two pieces to be joined are heated upto .... point to form the weld.
2. In .... welding heat is generated by introducing arc between electrode and workpiece.
3. The pressure required to hold the pieces together while current flows through the metal piece is called .... pressure.
4. In .... weld, the end to be joined must be squared.
5. The joint obtained in .... welding is much stronger than butt welding.
6. .... metal is used as additional metal to form the weld by melting the wire.
7. .... is used to prevent the oxidation of surface.
8. Small depression, called..........is formed in workpiece and molten metal is deposited around the edge of depression.
9. .... provide protective atmosphere, improve stability of arc and remove impurities.
10. In ....polarity electrode forms the negative terminal and workpiece forms the positive terminal.
11. The weight of coating in coating electrode is .... % of weight of electrode rod.
12. Acetylene cylinder is filled with spongy material saturated with chemical solvent known as ....
13. Soft solder is an alloy of .... and ....
14. Brazing is a process in which metal pieces are joined by means of filler material called ....
15. In .... welding polarity is not fixed at terminal.

C. Multiple Choice Questions

1. Joining of two pieces by means of rivet is
   (a) Temporary fastening
   (b) Semi permanent fastening
   (c) Permanent fastening
   (d) none
2. Two pieces to be joined are heated in furnace to fusion temperature and pressure is applied by the use of hammer in
   (a) Fusion welding (b) Thermit welding
   (c) Forge welding (d) Electric arc welding
3. Welding in which two pieces are welded by chemical action of aluminium and iron oxide is
   (a) Fusion welding (b) Thermit welding
   (c) Forge welding (d) Electric arc welding
4. Time period during which current flows to raise the temperature is
   (a) Squeeze time (b) Hold time
   (c) Weld time (d) Off time **[UPTU Dec. 2010]**
5. Two overlapping sheets are joined by application of pressure at one point
   (a) Resistance seam welding (b) Percussion welding
   (c) Resistance spot welding (d) Flash welding
6. Circular electrodes are used to form the weld in
   (a) Resistance seam welding
   (b) Percussion welding
   (c) Resistance spot welding
   (d) Flash welding
7. Which process is also called electrostatic energy process?
   (a) Resistance seam welding (b) Percussion welding
   (c) Resistance spot welding (d) Flash welding
8. Welding in which only D.C with straight polarity is used
   (a) Carbon arc welding (b) Resistance welding
   (c) Gas welding (d) none
9. Deflection of arc from its path due to generation of magnetic field in D.C. welding is called
   (a) Arc crater (b) Spatter
   (c) Arc blow (d) None
10. Flux
    (a) Provide protective atmosphere
    (b) Improve stability
    (c) Acts as deoxidizer
    (d) All of the above.
11. Gas welding is .... process.
    (a) Forge welding (b) Fusion welding
    (c) Pressure welding (d) None

12. Which flame is produced when supply of acetylene is higher than supply of oxygen.
(a) Carburizing flame (b) Oxidizing flame
(c) Neutral flame (d) None

13. Which metal is welded by using oxidizing flame?
(a) Brass (b) Aluminium
(c) Lead (d) Stainless steel.

14. For welding of lead which flame is used.
(a) Carburizing flame (b) Neutral flame
(c) Oxidizing flame (d) None

15. Hard solder is an alloy of
(a) Lead and tin (b) Copper and zinc
(c) Copper and lead (d) Zinc and tin

16. Welding is
(a) Temporary fastening (b) Semipermanent fastening
(c) Permanent fastening (d) None of these

17. Process of joining two or more elements is
(a) Forming (b) Fabrication
(c) Casting (d) Machining

18. Which welding is not type of electric arc welding
(a) Thermit welding (b) Metal arc welding
(c) Carbon arc welding (d) Tungsten arc welding

19. Process of joining two metal pieces by application of mechanical pressure and heat due to resistance offered by two pieces to flow of current
(a) Fusion welding (b) Forge welding
(c) Resistance welding (d) Soldering

20. Joint formed when ends of metal pieces are joined
(a) Lap joint (b) Butt joint
(c) Corner joint (d) T joint

21. Which joint is used for plates having thickness less than 3 mm
(a) Lap joint (b) Butt joint
(c) Corner joint (d) T joint

22. Which joint is economical due to less requirement of edge preparation
(a) Lap joint (b) T joint
(c) Corner joint (d) Edge joint

23. In which welding flux is not used
(a) Electric arc welding (b) Resistance welding
(c) Gas welding (d) Electroslag welding

24. The pressure required to hold the piece together while current flows through them
(a) Weld pressure (b) Forge pressure
(c) Hold pressure (d) Gap pressure

25. Time period during which forge pressure is applied to squeeze the metal pieces together in resistance welding

(a) Squeeze time (b) Off time

(c) Hold time (d) Weld time

26. Weld size is determined by ______ of electrode

(a) Length (b) Contact area

(c) Volume (d) All of these

27. What is the range of current in resistance welding

(a) 100 to 1000 A (b) 1000 to 3000 A

(c) 3000 to 100,000 A (d) 100,000 to 300,000 A

28. Electrodes used in resistance spot welding are made up of ______

(a) Aluminium base alloys (b) Silver base alloys

(c) Copper base alloys (d) Zinc base alloys

29. The resistance welding in which bars or rods of same cross-section are joined together with their ends facing each other

(a) Spot welding (b) Butt welding

(c) Seam welding (d) Projection welding

30. The resistance welding in which continuous circular rotating electrodes are used

(a) Spot welding (b) Projection welding

(c) Flash welding (d) Seam welding

31. Which welding is also defined as continuous series of spot welds

(a) Seam welding (b) Projection welding

(c) Flash welding (d) Spot welding

32. The welding in which heat is obtained by an electric arc produced by rapid discharge of stored electrostatic energy

(a) Seam welding (b) Projection welding

(c) Percussion welding (d) Spot welding

33. Percussion welding is limited to welding of ______ joints only

(a) Butt (b) Lap

(c) T (d) Corner

34. Which welding enables the production of several spot welds simultaneously

(a) Spot welding (b) Projection welding

(c) Flash welding (d) Percussion welding

35. Additional material is added to form the weld by melting the wire known as

(a) Filler metal (b) Flux

(c) Electrode (d) Arc blow

36. What type of polarity is used in carbon are welding

(a) AC with straight polarity (b) AC with reverse polarity

(c) DC with straight polarity (d) DC with reverse polarity

37. Which electrode has same composition as that of parent metal

(a) Coated electrode (b) Bare electrode

(c) Both (d) None of these

38. Small depression formed due to blast of arc is called

(a) Arc blow (b) Flux

(c) Arc crater (d) None of these

39. Metal arc welding is suitable for welding of plate having thickness

(a) 1 to 3 mm (b) 3 to 20 mm

(c) 25 to 45 mm (d) 50 to 70 mm

40. The deflection of arc from its path is called

(a) Arc blow (b) Arc crater

(c) Flux (d) None of these

41. Which power source is used in arc welding

(a) AC (b) DC

(c) Both AC and DC (d) None of these

42. In DC welding heat developed at positive terminal is _______ of total heat.

(a) 1/4th (b) 2/3rd

(c) 1/3rd (d) 3/4th

43. In which type of welding, heat generated at each pole is same

(a) AC (b) DC

(c) Both (d) None of these

44. The polarity in which electrode forms the negative terminal and workpiece forms the positive terminal

(a) Reverse (b) Straight

(c) Both (d) None of these

45. Which polarity is used is welding of thin materials

(a) Reverse (b) Straight

(c) Both (d) None of these

46. In which welding problem of arc blow exists

(a) AC (b) DC

(c) Both (d) None of these

47. Which electrode is used for AC and DC welding

(a) Graphite (b) Carbon

(c) Tungsten (d) None of these

48. Weight of coating on coated electrode is _______ of weight of electrode rod

(a) 1.5 to 30% (b) 30 to 40%

(c) 40 to 50% (d) 45 to 50%

49. _______ is used to dissolve oxide and impurities present in the surface of metal

(a) Filler (b) Flux

(c) Electrode (d) All of these

50. The fusion welding process that joins the metal by using heat of combustion of strong gas flame.

(a) Electric arc welding (b) Resistance welding

(c) Gas welding (d) Spot welding

51. Oxygen gas cylinder is painted with ______ colour

(a) Maroon (b) Green

(c) Red (d) Black

52. Acetylene gas cylinder is filled with spongy material saturated with chemical solvent known as

(a) Methane (b) Propane

(c) Acetone (d) Alcohol

53. Which component of oxy-acetylene welding is used to produce a steady flow of gas under varying pressure ?

(a) Hose pipe (b) Blow pipe

(c) Pressure regulator (d) Stop valve

54. ______ flame doesn't create any chemical change in the molten metal ?

(a) Neutral flame (b) Oxidizing flame

(c) Carburizing flame (d) None of these

55. ______ flame is used for welding of cast iron, aluminium and stainless steel

(a) Oxidizing flame (b) Neutral flame

(c) Carburizing flame (d) None of these

56. Carburizing flame of maximum temperature is **[UPTU Dec. 2010]**

(a) 1500°C (b) 2500°C

(c) 3038°C (d) 4000°C

57. Which flame is produced when supply of oxygen is higher than supply of acetylene.

(a) Carburizing flame (b) Neutral flame

(c) Oxidizing flame (d) None of these

58. Method of joining sheet metals by using filler metals whose liquidus temperature is below 427°C is called

(a) Brazing (b) Soldering

(c) Gas welding (d) Resistance welding

59. Soft solder is an alloy of

(a) Copper and tin (b) Copper and lead

(c) Tin and lead (d) Copper and silver

60. Filler metal used in brazing is mixture of

(a) Cu and Sn (b) Cu and Pb

(c) Sn and Zn (d) Sn, Cu and Zn

61. In brazing process liquidus temperature of filler metal is

(a) Below 427°C (b) Above 427°C

(c) Equal to 427°C (d) None of these

D. Match the Following

1.

(A)	(B)
(i) Arc crater	(a) Act as deoxidizer
(ii) Arc blow	(b) Small depression formed in workpiece
(iii) Flux	(c) Additional material to form the meld.
(iv) Filler metal	(d) Deflection of arc from its path

2.

	(A)		(B)
(i)	Circular electrodes	(a)	Projection welding
(ii)	Continuous series of spot weld	(b)	Resistance butt welding
(iii)	Electrostatic energy process	(c)	Seam welding
(iv)	Same crosssection of workpiece	(d)	Percussion welding

3.

	(A)		(B)
(i)	Forge welding	(a)	Application of mechanical pressure and heat
(ii)	Oxyacetylene welding	(b)	Stream of electrons
(iii)	Resistance welding	(c)	Use of hammer
(iv)	Electric arc welding	(d)	Mixure of oxygen and acetylene

4.

	(A)		(B)
(i)	Straight polarity	(a)	Electrode positive, workpiece negative
(ii)	Reverse polarity	(b)	Alloy of tin and lead
(iii)	Soft solder	(c)	Alloy of Cu and Zinc
(iv)	Hard solder	(d)	Electrode negative, workpiece positive

ANSWERS

A. True or False

1. False **2.** False **3.** True **4.** False **5.** False
6. False **7.** True **8.** False **9.** False **10.** True
11. False **12.** False

B. Fill in the Blanks

1. Fusion
2. Electric Arc
3. Weld
4. Butt
5. Flash
6. Filler
7. Flux
8. Arc Crater
9. Flux
10. Straight
11. 1.5 to 30%
12. Acetone
13. Lead and Tin
14. Spelter
15. A.C

C. Multiple Choice Questions

1. (b) **2.** (c) **3.** (b) **4.** (c) **5.** (c)
6. (a) **7.** (b) **8.** (a) **9.** (c) **10.** (d)
11. (b) **12.** (a) **13.** (a) **14.** (a) **15.** (b)
16. (c) **17.** (b) **18.** (a) **19.** (c) **20.** (b)
21. (a) **22.** (d) **23.** (b) **24.** (a) **25.** (a)
26. (b) **27.** (c) **28.** (c) **29.** (b) **30.** (d)
31. (a) **32.** (c) **33.** (a) **34.** (b) **35.** (a)
36. (c) **37.** (b) **38.** (c) **39.** (b) **40.** (a)
41. (c) **42.** (b) **43.** (a) **44.** (b) **45.** (a)

46. (b)	**47.** (c)	**48.** (a)	**49.** (b)	**50.** (c)
51. (d)	**52.** (c)	**53.** (c)	**54.** (a)	**55.** (b)
56. (c)	**57.** (c)	**58.** (b)	**59.** (c)	**60.** (d)
61. (b)				

D. Match the Following

1. (i)—(b), (ii)—(d), (iii)—(a), (iv)—(c) **2.** (i)—(c), (ii)—(a), (iii)—(d), (iv)—(b)

3. (i)—(c), (ii)—(d), (iii)—(a), (iv)—(b) **4.** (i)—(d), (ii)—(a), (iii)—(b), (iv)—(c)

❑❑❑

UNIT-V

Miscellaneous Topics

9. MANUFACTURING AND PLANT LAYOUT

10. POWDER METALLURGY AND PLASTICS

SYLLABUS

Manufacturing : Importance of Materials and Manufacturing Towards Technological and Socio-Economic Developments. Plant Location. Plant Layout – its Types. Types of Production. Production Versus Productivity.

Misc. Processes : Powder-Metallurgy Process and its Applications, Plastic-Products Manufacturing, Galvanizing and Electroplating.

CHAPTER 9

Manufacturing and Plant Layout

INSIDE THIS CHAPTER

9.1. IMPORTANCE OF MATERIALS FOR TECHNOLOGICAL AND SOCIO-ECONOMIC DEVELOPMENTS

[UPTU DEC. 2008; MAY 2009]

Material plays a pivotal role in our culture since pre-civilization era. The role of material in civilization is so crucial that the civilizations are named by the material such as stone age, bronze age, steel age, etc. In present time materials are still important to improve economic performance and the quality of life in the following main areas :

1. Environment

Environment has prime importance due to its direct impact on human beings. Due to high polluting environment, technology is changing to prevent the pollution. Material science is helping to discover new energy generation technology, less toxic and easily recyclable materials, etc.

2. Health

Materials are now meeting the challenges to provide solution for better health by providing water filtration systems, safe drugs, organ implants, artificial bones, etc.

3. Consumer Products

Present era is the era of quality, so to provide better consumer products at low cost new materials are used. These new materials are resulting in faster production and better quality goods. Materials are now easily fabricated, machined, casted, etc. due to advances in material technology.

4. Communication

Everybody is aware of the need of faster communication in the present time, so the developments are on the way to find out new electronic, magnetic and optical materials to fulfill the need.

5. Production

Material with required properties can now easily be fabricated, machined, welded etc. to perform their function satisfactorily thus satisfying the service and economic requirements of industry.

9.2. IMPORTANCE OF MANUFACTURING IN TECHNOLOGICAL AND SOCIO-ECONOMIC DEVELOPMENTS

[UPTU Dec. 2008; May 2009]

Manufacturing is an indicator of the growth and prosperity of a nation. The rate of manufacturing is consistently correlated with fluctuations in the economy of nations. Manufacturing not only imparts value to the raw material by converting it into desired product but it also imparts value to the nation by creation of wealth and better standard of living.

Manufacturing is the transformation of raw materials into finished goods by various processes and operations. For the product to be economically viable in present competitive market, it should have least cost and better quality. To achieve this, work should be done at exact dimensions, scrap should be minimum, production rate should be high, inprocess inventory should be less, etc.

All these aspects lead to advancement in technology by using CNC machines, better layout, inventory planning etc. Other advantages of manufacturing are :

1. Developments of decision making power in employees.
2. Develops team spirit in management and workers.
3. Mutual trust and integrity in management and workers.

9.3. PLANT LOCATION

A plant is a place where men, machines, materials, equipments are arranged in a systematic way. Plant location means selecting a proper place for plant where it can produce the parts at minimum cost to face the market competition. Various factors considered in plant location are :

1. Availability and Nearness of Raw Material

Raw material is the prime requirement of industry. It should be easily and economically available in abundance round the year. Large distance of plant from the source of raw material increases the transportation cost so plant should be located within economic distances.

2. Nearness of Market

Plant should be located near to market to reduce the transportation cost and avoid damaging of finished product. Due to small distance, plant can acquire large part of market and can provide fast service to customers.

3. Transportation Facilities

From the above two factors it is evident that transportation facility is one of the major factor for selecting the location. So while selecting a location, transportation cost should be minimized by checking the availability of a proper and cheap transportation medium near to plant to cut down the total cost.

4. Availability of Workers

Sufficient workers at reasonable rate and with right attitude towards work must be available near to plant to achieve lower labour cost per unit of production.

5. Availability of Electricity or Fuel

Plant should be located near to source of fuel to cut down the cost of fuel transportation. Electricity should also be available continuously and at reasonable rates.

6. Cost of Land

Sufficient land should be available at low cost with low probability of flood and earthquakes.

7. Climate

Climate condition should be good and favourable for human beings. Also climate should meet the requirement of particular industry.

8. Other Factors

It includes law and order, taxes, housing facility, hospitals, markets, schools, banks, social infrastructure, security facilities, presence of related industries, availability of water, proper drainage, etc.

9.4. PLANT LAYOUT [UPTU DEC. 2008]

Plant layout means an optimum arrangement of various department, manufacturing shops, machines, equipments and services of an organization. A good plant layout help at the processing of product in the most efficient and economic manner in least possible time. The main aim of good plant layout is to utilize all available resources to optimize production and increase productivity.

9.4.1. Objectives of Plant Layout

1. To minimize material movement.
2. To reduce inprocess inventory.
3. To smooth the material flow.
4. To minimize movement of workers.
5. To obtain higher productivity.
6. To effectively utilize, man, machine and space.
7. To provide flexibility in operations.
8. To provide safe working condition and satisfaction to workers.
9. To avoid unnecessary investment.
10. To provide better quality of product at low cost.
11. To provide better ventilation, lighting etc.
12. To utilize cubic space.

9.4.2. Advantages of Good Plant Layout

1. Effective utilization of man, machine and other resources.
2. Enhancement of overall efficiency.
3. Better supervision and control of plant.
4. Easy maintenance of plant and machinery.
5. Reduced production time.
6. Minimum material handling time.
7. Safe and good working conditions.
8. Increased rate of production.
9. Proper utilization of floor area.
10. Minimum indirect expenses.

9.4.3. Factors Affecting Plant Layout

1. Use of material handling facilities.
2. Different processes of manufacturing.
3. Type of production process.
4. Size and specification of material and product.
5. Policy of management regarding employee facilities.
6. Shifting of equipment to other place.
7. Introduction of products in future.
8. Storage facilities.

9.5. TYPES OF LAYOUTS

9.5.1. Line Layout or Product Layout

In this type of layout all manufacturing facilities are arranged in a sequence in which they are required to produce a product. It is ideal form of layout. In this layout material flows in a forward direction along a prescribed line from one machine to another and leave the last machine when manufacturing work is complete. It is a continuous production involving a continuous flow of material from raw material stage to finished stage.

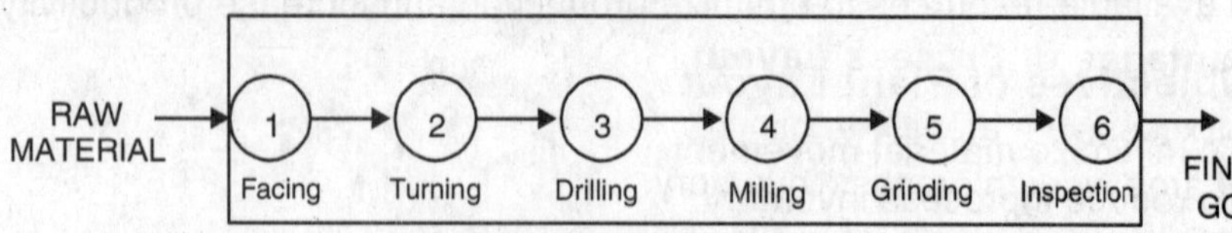

Fig. 9.1. Line Layout.

Advantages of Line Layout

1. Work flow is continuous and smooth.
2. Less material handling is involved.
3. Efficient and effective supervision.
4. Less floor area is required.
5. Less time in manufacturing of product.
6. Optimum utilization of machines and equipments.

Disadvantages of Line Layout

1. Difficult to change the product design.
2. High capital investment.
3. Breakdown of single work station in a production line stops the whole production.

9.5.2. Process Layout

Process layout is also called as functional layout and is used for batch production. In this type of layout similar types of machines and equipment are installed at one place so that all the operations of same type are performed at one place e.g all lathe machine are installed at one place, called lathe machine shop, all milling machine are installed at another place, called milling shop.

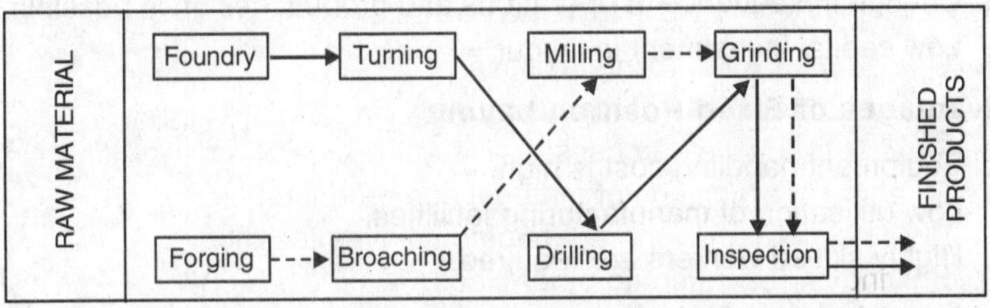

Fig. 9.2. Process Layout.

Advantages of Process Layout

1. Machines and equipments are better utilized.
2. Workers are satisfied due to different department for different processes as they can develop expertise in working or supervision of one type of work.
3. All shops of plant can run independently.
4. Health hazard in one section do not affect the worker of other section.
5. Less capital investment.
6. Greater flexibility regarding distribution of work to machines and workers.

Disadvantages of Process Layout

1. Excessive material handling.
2. Large work in process inventory.
3. High supervision cost.
4. High labour cost.
5. Production planning and control is relatively difficult.

9.5.3. Fixed Position Layout

In this layout workpiece is fixed at one place and men, machines and materials are moved to it to work upon. The location of major component or material is not disturbed till the product is completely manufactured or assembled. It is preferred for the jobs which are dimensionally big such as ship, building, aeroplanes, and flyover.

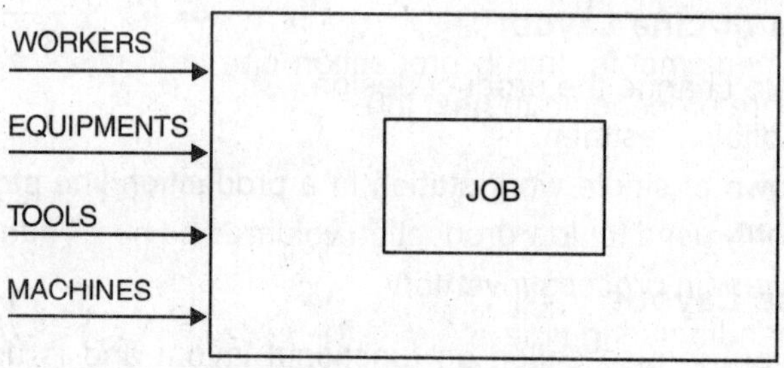

Fig. 9.3. Fixed Position Layout.

Advantages of Fixed Position Layout

1. Least movement of material and workers.
2. Change in sequence of operations and product design is possible.
3. Low capital investment in layout.

Disadvantages of Fixed Position Layout

1. Equipment handling cost is high.
2. Low utilisation of manufacturing facilities.
3. Highly skilled workers are required.

9.5.4. Combination Layout

In industries, only one form of layout is rarely used. To obtain the best result layout is designed by combining all desirable features of product and process layout into a single one known as combination layout. In this type of layout a set of equipment or machinery is grouped together in a section and this group of similar machines is then arranged according to sequence of manufacturing, so that each set of machines or equipments is used to perform similar operation to produce a family of component. This type of layout is also known as group layout.

Advantages of Combined Layout

1. Shorter product lead time.
2. Best utilization of man, material and machinery.
3. Better quality and low material handling cost.

9.6. PRODUCTION [UPTU DEC. 2008]

It is defined as a process which converts raw materials into finished products with the application of tools and techniques. Production refers to absolute output i.e. the number of parts produced.

9.6.1 Types of Production [UPTU May 2009]

1. Job Shop Production
2. Batch Production
3. Mass Production

1. Job Shop Production

It is characterized by low production volume and high variety. It is used to produce the product as per customer specification by using general purpose machines and equipments. In job production operators work upon a job and complete it before proceeding to next job.

Features

1. Commonly used for low production volume and high variety of products.
2. Large work in process inventory.
3. High manufacturing time.
4. Highly skilled workers are required.

2. Batch Production

Batch production is used for production of medium volume lots and is suitable for moderate variety of products. Examples of batch production are furniture making, manufacturing of medicine, etc.

In batch production an operation is completed on each of the identical item before the next operation is performed.

Features

1. Less manufacturing time as compared to job shop production.
2. Large work in process inventory.
3. Good flexibility.
4. Skilled labour with specialization in one field.
5. General purpose equipments are used.

3. Mass Production

Mass production is used to produce low variety of parts having high volume and is characterized by high rate of production. Examples are nuts, bolts, automobiles, etc. In mass production, each of the identical item is passed for second operation as soon as first operation is completed and so on till the item is completely manufactured.

Features

1. It has little or no flexibility.
2. Manufacturing time is very short.
3. Special purpose machines and tools are used.
4. Less skilled workers are required.
5. Assembly line has a product layout.
6. Product flow is straight and predictable.

9.7. PRODUCTIVITY [UPTU DEC. 2008]

It is generally defined as ratio of output and input in any system. Productivity has different meanings in different prospects. Output generally means amount produced or number of items produced and input means various resources such as equipment, machinery, material, labour, etc. used to produce the output.

$$\text{Productivity} = \frac{\text{Output}}{\text{Input}}$$

Increase in output is not an indicator of increase in productivity but the increase in output for same input indicates the increase of productivity. High productivity decreases the cost of production and thus decreases cost of product, making product available at cheaper rate to customer. High value of productivity also leads to better wages and improved working condition due to increase in profit of company.

Various factors which affect productivity are :

1. Human resources.
2. Product or system design.
3. Production volume.
4. Technology and capital investment.
5. Machinery and equipment.
6. Government regulation.
7. Skill and effectiveness of worker.

9.8. PRODUCTION Vs. PRODUCTIVITY

1. Production is related to conversion of raw material into finished products whereas productivity is related to how efficiently the conversion is done.
2. Production is a measure of produced output whereas productivity is measure of effective utilization of input.
3. Increase in output results in increase of production whereas increase in output with same input results in increase in productivity.
4. Production is an absolute concept whereas productivity is a relative concept.

SUMMARY

- A plant is a place where men, machines, materials, equipments are arranged in a systematic way.
- Plant location means selecting a proper place for plant where it can produce the parts at minimum cost.
- Plant layout means an optimum arrangement of various department, manufacturing shops, machines, equipments and services of an organization.
- In line layout all manufacturing facilities are arranged in a sequence in which they are required to produce a product.
- In process layout similar type of machines and equipments are installed at one place so that all the operations of same type are performed at one place.
- In fixed position layout workpiece is fixed at one place and men, machines and materials are moved to it to work upon.
- Production is defined as a process which converts raw materials into finished products with the application of tools and techniques.
- Job shop production is used to produce the part as per customer specification by using general purpose machines and equipments.

- Batch production is used for production of medium volume lots and is suitable for moderate variety of products.
- Mass production is used to produce low variety of parts having high volume.

SHORT QUESTIONS WITH ANSWERS

Q. 1. What do you mean by plant?
Ans. A plant is a place where men, machine, material and equipment are arranged in a systematic way.
Q. 2. What is meant by plant location?
Ans. Plant location means selecting a proper place for plant where it can produce parts at minimum cost.
Q. 3. List four major factors in plant location.
Ans. (1) Availability of raw material
(2) Nearness of market
(3) Transportation facilities
(4) Availability of workers
Q. 4. What is plant layout?
Ans. Plant layout means an optimum arrangement of various departments, manufacturing shops, equipments and services of an organization.
Q. 5. What is the main aim of plant layout?
Ans. The main aim of plant layout is to utilize all available resources to optimize production and increase productivity.
Q. 6. What are different types of layouts?
(a) Line layout or Product layout
(b) Process layout
(c) Fixed position layout
(d) Combination layout
Q. 7. Which layout is suitable for batch production?
Ans. Process layout is suitable for batch production.
Q. 8. What are different types of production?
Ans. (a) Job shop production
(b) Batch production
(c) Mass production
Q. 9. Give two examples of fixed position layout. **[UPTU May 2012]**
Ans. Fixed position layout used in ship and aeroplanes.

REVIEW QUESTIONS

1. Discuss importance of materials towards technological and socio-economic considerations.
2. Discuss importance of manufacturing towards technological and socio-economic considerations.
3. What do you mean by plant layout? Discuss objectives and advantages of plant layout. **[UPTU Dec. 2008]**
4. Explain different types of plant layout with their advantages and disadvantages and applications. **[UPTU Dec. 2009]**
5. What is the difference between production and productivity? **[UPTU May 2012]**
6. Explain different types of production.

7. Define plant location. Explain various factors that are considered for plant location.
8. Discuss the role and importance of materials and manufacturing for growth of any nation. Explain production and productivity.
[UPTU Dec. 2008; May 2010]
9. Discuss product layout and process layout with suitable examples.
[UPTU May 2012]

SELF TEST QUESTIONS

A. State Whether True or False

1. Product layout is suitable for automobile manufacturing plant.
2. Fixed position layout is suitable for ship vessel industry.
3. In job shop production product variety is low.
4. Increase in output, without increase in input means increase in productivity.
5. Product layout has high flexibility.

B. Fill in The Blanks

1. Plant location means selecting a proper .... for plant.
2. Good plant layout consists of effective utilization of ....
3. Disadvantages of process layout are.......... work-in process inventory and .... supervision cost.
4. Combination layout is also known as ....
5. Batch production has less manufacturing time as compared to ....
6. Manufacturing is the process of converting raw material into ....
[UPTU Dec. 2008]

C. Multiple Choice Questions

1. Product layout is used for
(a) Continuous production (b) Batch production
(c) Job shop production (d) All of the above.
2. Process layout is used for
(a) Continuous production (b) Batch production
(c) Job shop production (d) All of the above.
3. Fixed position layout is used for
(a) Continuous production (b) Batch production
(c) Job shop production (d) All of the above.
4. In which type of production breakdown of single work station stops the whole production.
(a) Process layout (b) Product layout
(c) Fixed position layout (d) Combination layout.
5. Batch production is suitable for
(a) High production volume (b) Medium production volume
(c) Low production volume (d) All of the above.
6. Similar type of machine and equipments are installed at one place in
(a) Process layout (b) Product layout
(c) Fixed position layout (d) Combination layout
7. Combination layout combines the features of **[UPTU Dec. 2010]**
(a) Product and process layout (b) Job shop and product layout
(c) Job shop and process layout (d) None

8. Productivity is measure of
(a) Produced output (b) Effective utilization of input.
(c) Input Vs output (d) Human resources.

9. Place where men, machines, materials equipments are arranged in a systematic way.
(a) Plant location (b) Plant
(c) Plant layout (d) Production

10. Selection of proper place for plant where it can produce the parts of minimum cost
(a) Plant (b) Layout
(c) Plant location (d) Production

11. Large distance of plant from the source of raw materials increases
(a) Production cost (b) Machining Cost
(c) Marketing cost (d) Transportation cost

12. Factors considered in plant location are
(a) Nearness of market (b) Availability of workers
(c) Cost of land (d) All of these

13. Optimum arrangement of various departments, manufacturing shops, machines, equipments and services of an organization is known as
(a) Plant (b) Production
(c) Plant location (d) Plant layout

14. Layout in which manufacturing facilities are arranged in a squence in which they are required to produce a product.
(a) Line layout (b) Process layout
(c) Fixed position layout (d) Combination layout

15. Which layout is also called as functional layout and used for batch production ?
(a) Line layout (b) Process layout
(c) Fixed position layout (d) Combination layout

16. In which type of layout all lathe machine are installed at one place
(a) Line layout (b) Fixed position layout
(c) Process layout (d) Combination layout

17. Which layout is preferred for dimensionally big jobs such as ship, building, aeroplanes
(a) Line layout (b) Fixed position layout
(c) Process layout (d) Combination layout

18. Process which converts raw materials into finished products with the application of tools and techniques
(a) Plant layout (b) Plant
(c) Production (d) Plant location

19. ______ is characterized by low production volume and high variety
(a) Job shop production (b) Batch production
(c) Mass production (d) None of these

20. Ratio of output and input gives
(a) Production (b) Productivity
(c) Effectiveness (d) All of these

21. _______ is characterized by high production volume and low variety
(a) Job shop production (b) Batch production
(c) Mass production (d) None of these

22. High productivity _____ cost of product
(a) Increases (b) Decreases
(c) Not effective (d) None of these

23. Production is a measure of
(a) Produced output (b) Effective utilization of input
(c) Input vs output (d) Human resources

D. Match the Following

1.

	(A)		(B)
(i)	Job shop production	(a)	Producton of medium volume lots
(ii)	Batch production	(b)	Production of high volume lots with low variety
(iii)	Mass production	(c)	Ratio of output and input
(iv)	Productivity	(d)	Production of low volume lots with high variety

2.

	(A) Type of Products		(B) Layout
(i)	Ball bearings	(a)	Process layout
(ii)	Tools and gauges	(b)	Product layout
(iii)	Motor cycle assembly	(c)	Fixed position layout
(iv)	Large boilers	(d)	Combination of product and process layout

ANSWERS

A. True or False

1. True **2.** True **3.** False **4.** True **5.** False

B. Fill in the Blanks

1. Place **2.** Resources **3.** Large, high **4.** Group layout
5. Job shop production 6. Finished products

C. Multiple Choice Questions

1. (a)	**2.** (b)	**3.** (c)	**4.** (b)	**5.** (b)
6. (a)	**7.** (a)	**8.** (b)	**9.** (b)	**10.** (c)
11. (d)	**12.** (d)	**13.** (d)	**14.** (a)	**15.** (b)
16. (c)	**17.** (b)	**18.** (c)	**19.** (a)	**20.** (b)
21. (c)	**22.** (b)	**23.** (a)		

D. Match the Following

1. (i)—(d), (ii)—(a), (iii)—(b), (iv)—(c) **2.** (i)—(d), (ii)—(a), (iii)—(b), (iv)—(c)

❑❑❑

CHAPTER 10

Powder Metallurgy and Plastics

INSIDE THIS CHAPTER

10.1. INTRODUCTION

Powder metallurgy is an art of manufacturing commercial articles from powdered metals and alloys. This process consists of blending and pressing (compacting) of powdered particles into desired shape and then heating in a controlled atmosphere to bind the particles and improve their properties (sintering).

- This process is economically feasible for mass production of small, intricate parts of high precision and also eliminates the need for additional machining or finishing.
- It can produce controlled degree of porosity.
- It is used where desired properties and characteristic are difficult to obtain by any other method.
- Automobile applications accounts for nearly 75% of powder metallurgy market.
- Some examples of powder metallurgy products are oilless self-lubricating bearings, tungsten carbide cutting tools, electrical contacts, turbine blades, copper coins, tungsten wire, household appliances, etc.

10.2. POWDER METALLURGY PROCESS

Powder metallurgy consists of following stages :

1. Powder Manufacture
2. Mixing or Blending
3. Compacting
4. Sintering

10.3. POWDER MANUFACTURE

Properties of powder metallurgy product depends upon the properties of metal powder such as particle shape, particle size, size distribution and surface texture of particles. Following processes are commonly used for producing powders.

1. Atomization

In this process molten metal is fragmented into droplets by using stream of water and air. These metal droplets solidifies into small particles of various irregular shapes.

Procedure : Molten metal is filled in hopper like container called tundish, which is fitted on top of container. As metal flows down in form of stream through the nozzle, water strikes from both sides on stream of metal, causing rapid cooling of metal in form of small particles. The size and shape of particles depends upon velocity and medium of atomizing jets, temperature of liquid, and environment provided for cooling.

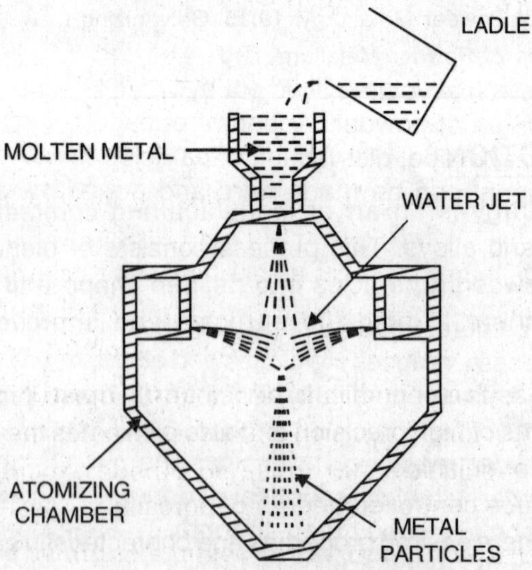

Fig. 10.1. Atomization.

Another technique of atomization is rotating electrode process which consists of two electrodes, one made of tungsten and other made of metal whose atomization is to be done. Tungsten electrode is nonrotating and nonconsumable whereas other electrode is rotating and consumable. Arc is strucked between the electrodes due to which intense heat is produced and the metal melts from rotating electrodes. Molten metal is cooled upto desired temperature and get converted into small particles which are collected at the bottom.

2. Reduction Process

In this process oxides or refined ores are reduced in presence of reducing agent at a temperature below their melting point by chemical process. This reducing agent may be carbon monoxide or hydrogen. The spongy material, obtained after reduction is crushed and ground to produce a powder.

3. Electrolysis

This process is similar to electroplating in which one metal plate acts as anode and other metal plate act as cathode. When high current is passed through the electrolyte, metal powder is deposited on cathode. Cathode is washed with water and dried. After drying it is scrapped off and pulverized to produce powder of desired grain size. This process produces the fine powder and is used for tin, silver, and copper metals.

4. Milling or Crushing

In this process millers, crushers and stampers break down the metal into small particle by crushing and impact. This process is suitable for brittle material.

5. Shotting

In this process molten metal is passed through a sieve into water to produce large spherical particles.

10.4. PROPERTIES OF METAL POWDER

1. Apparent Density

It is defined as actual volume filled out by loose powder and represented as mass per unit volume of powder. Apparent density depends upon number of factors such as size, shape, distribution of particles and chemical composition. Portion of component can be made hard and dense by controlling apparent density.

2. Purity

Powder must be pure. Impurity present in the metal powder affects sintering and briquetting process, reduces life of dies and also lower mechanical properties.

3. Particle Shape

Method of powder manufacture decides the particle shape which may be flat, irregular, angular, lamellar, spherical or dendritic. In these shapes irregular shaped particles achieve good green strength. The shape of the particle affects compactness, strength and porosity.

4. Size of Particle and Distribution

The size of particle is defined by many methods. It has good influence over porosity, density, compressibility and shrinkage. If size of particles is same then porosity of metal increases otherwise decreases. If particles have fine grains, density of metal increases otherwise decreases.

5. Flowability

It shows the flowing ability of metal powder. Time taken by a definite quantity of powder to flow out of a standard orifice is known as flow rate. There are two parameters that are governed by flow rate one is production rate and other is time required to fill the mould. Flowability depends upon size, shape and distribution of particles.

6. Compressibility

It is defined as the ratio of initial volume of the powder (unpressed volume) to final volume (pressed volume) after compression. It imparts green strength to powder also known as mechanical strength of compressed part.

7. Sintering Ability

It is the ability by which metal powder promotes adhesion between particles. It should be high enough to produce good quality products. Regular shapes have good sintering properties.

10.5. PROCESSING OF POWDER

Processing of powder has been classified into two groups.

1. Primary processes
2. Secondary processes

10.5.1. Primary Process

(i) Blending or Mixing
(ii) Briquetting or Compacting
(iii) Presintering
(iv) Sintering

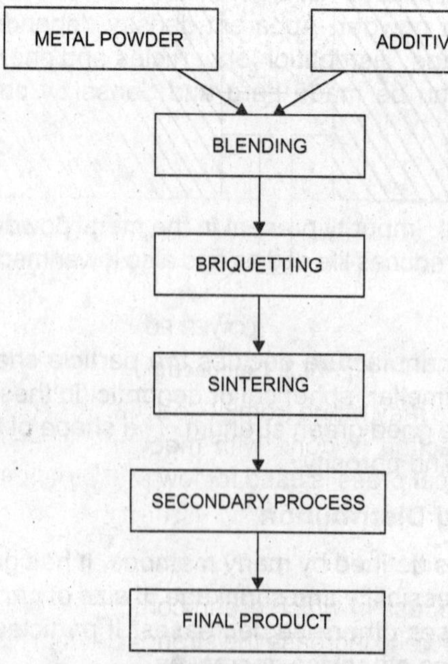

Fig. 10.2. Flow Chart of Powder Metallurgy Process.

(i) Blending or Mixing

Blending is done to obtain uniform distribution of particle size. Blending is required when different metal powders are used or nonmetallic particles are added to impart some specific properties. Blending is not required for metal powder having correct particle size distribution. Mixing of constitutes may be wet or dry. In wet mixing, water or solvent is used to reduce explosion hazard and dusting. Lubricants such as graphite, stearic acid are added to reduce the die wall friction, friction between the particles and improve the flow characteristic, green strength and compressibility. Binders are also added to develop adequate green strength.

(ii) Briquetting or Compacting

In this process powders are compacted and densified into an accurate size or shape by application of pressure at room temperature and the resulting compact is known as green compact. Pressure increases the hardness, strength and density. Compacting is done in hardened steel dies by use of punches. If one punch is used, pressure will not be uniformly distributed throughout the mass, so two punches are used one from the top and other from the bottom. Lower punch is called ejector because it removes the compressed part. Powder is compressed nearly to one third of its original volume.

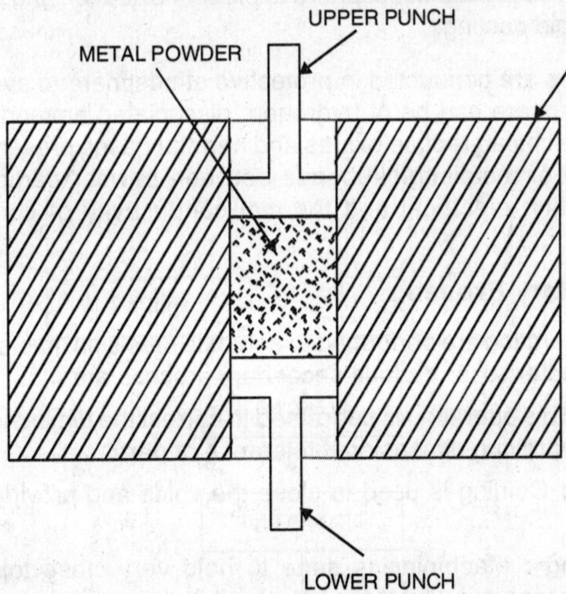

Fig. 10.3. Compacting.

Compacting pressure may range from 40 to 1650 MPa depending on material and application. Compacting is done with mechanical press, hydraulic press and rigid tools. Mechanical press is used for low pressure and hydraulic press is used for high pressure.

(iii) Presintering

It is defined as a process in which green compact is heated to temperature below sintering temperature to increase the strength. It also removes the lubricant and binders added during blending. Presintering is not done when machining is not required.

(iv) Sintering

In the sintering operation, compacted parts are heated to high temperature in controlled atmosphere but below the melting point of material being sintered. Heated part is held for certain period at correct temperature to permit bonding of particles. It increases strength and hardness of the part to the require extent. Most of metals are sintered up to certain temperature, which is 70% to 80% of their melting point. Some refractory metal are sintered upto 90%. If compacted part is made of more than one material then sintering temperature may be more than the melting temperature of one or more component.

Sintering operation involves three stages.

(a) Purge or Burn off Chamber : It is designed to remove volatile substances, lubricants, binders and to raise the temperature of compacts in controlled way.

(b) High Temperature Stage : To enable solid state diffusion and bonding between the particles, part is heated to sintering temperature and held for sufficient time varying from 10 minutes to several hours.

(c) Cooling Period : In this stage temperature of product is gradually lowered in controlled atmosphere to prevent oxidation and thermal shock due to rapid cooling.

All three states are conducted in protective atmosphere to avoid oxidation. Protective atmosphere can be of hydrogen, dissociated ammonia or cracked hydrocarbons, which can reduce oxides and harmful gases present in particles. Due to formation of metallurgical bonds between powder particles strength, toughness, ductility, electrical and thermal conductivity of compacted part increases.

10.5.2. Secondary Process

Secondary operation are performed to enhance precision, improve properties and special characteristics. Various secondary process are

(i) Sizing : This operation is performed to correct the dimension of sintered part by improving dimensional tolerance of part.

(ii) Coining : Coining is used to close the voids and provide the desired density.

(iii) Machining : Machining is done to hold very close tolerances and producing features which are not possible through primary processes.

(iv) Infiltration : In infiltration, molten metal is used to fill the porosity in the sintered compact. It increases the strength, density and hardness which is not possible through sintering alone.

(v) Impregnation : In this process pores in sintered part are filled with oil and provides self-lubricating properties to parts.

(vi) Plating : Plating of sintered part is done to prevent corrosion and provide protective atmosphere.

10.6. APPLICATIONS OF POWDER METALLURGY

1. Porous or permeable products such as bearings, filter and flow regulators.
2. Self lubricated bearings.
3. Clutch facing and brake lining.
4. Porous filters and bearings.
5. Magnets.
6. Cement carbide tools and dies.
7. Pump rotors and gears.
8. Motor bushes.

10.7. ADVANTAGES OF POWDER METALLURGY

1. Production rate is high.
2. Scrap is reduced.
3. Complex shapes can be made.
4. Controlled porosity is achieved.
5. Good surface finish and accurate dimension is obtained which eliminate further machining in most cases because of very high dimensional accuracy.
6. Labour cost is low.
7. Bimetallic and laminated parts can be produced.

10.8. DISADVANTAGES OF POWDER METALLURGY

1. Not economical for small scale production.
2. Dies and equipments cost is high.
3. Not possible to eliminate the porosity completely in parts.
4. Intricate designs are difficult to attain.
5. Size of part is limited.

10.9. PLASTICS

Plastics are synthetic organic material, which can be formed into different shapes with the application of heat and pressure. Plastics are characterized by large molecules called polymers which are composed of large number of small molecules called monomers. Synthetic resins are usually referred as plastics and their compounds can be moulded, extruded, cast or used as thin film or coatings. They offer low density, low tooling cost, good corrosion resistance, good dielectric strength and design versatility. Generally used natural polymers are resin, cellulose, shelter, etc.

10.10. PROPERTIES OF PLASTICS

1. Good thermal and electrical insulation capacity.
2. Light weight and low density.
3. Good surface finish is obtained.
4. High resistance to corrosion.

5. High antifriction properties.
6. Good chemical stability.
7. Not affected by gases and moisture.
8. Easy to mould due to low melting point.
9. High damping capacity.
10. High strength to weight ratio.

10.11. CLASSIFICATION OF PLASTICS

Plastics are classified into two main groups :

10.11.1. Thermosetting Plastics

These plastics can be soften only once and can't be resoften again under heat due to non-reversible chemical changes. In other words the useful properties of plastics get destroyed at high temperature making it unsuitable for reuse. These plastics have a highly cross-linked or three dimensional framework structures.

Example

(i) Phenol formaldehyde
(ii) Urea
(iii) Epoxy resins
(iv) Polyesters
(v) Alkyds
(vi) Epoxides or epoxy resins

10.11.2. Thermoplastic Plastics

These plastics can be repeatedly softened by heating and hardened by cooling so can be repeatedly shaped. Thermoplastic materials don't have a definite melting temperature. It softens over a range of temperature.

Example

(i) Polyethylene
(ii) Polystyrene
(iii) Vinyl resin
(iv) Polyamide
(v) Acrylic resin

10.11.3. Difference between Thermosetting and Thermoplastic

Thermosetting Plastic	Thermoplastic
1. Once set cannot be resoften and remoulded into any new shape by application of heat.	1. Can be resoften and remoulded into any shape by application of heat.
2. Formed by condensation polymerization.	2. Formed by addition polymerization.
3. Polymers of thermosetting plastic are composed of crosslinked molecules.	3. Polymers are composed of chain molecules.

Thermosetting Plastic	Thermoplastic
4. These plastics have better heat resistance, better chemical resistance and less creep.	4. These plastics have less resistance to heat and chemical effect and have high creep.
5. Cannot be mechanically softened or deformed at high temperature.	5. Can be mechanically deformed or softened at high temperature.
6. Application of heat leads to degradation and decomposition.	6. Application of heat soften the plastic and retain chemical composition.
7. These material become hard after heating (heat setting resins).	7. These material become hard after cooling (cold setting resins).
8. Making a part, rod, sheet or tube is difficult.	8. Various shapes can be produced easily.
9. Plasticity of these material does not increase with increase in temp-rature.	9. Plasticity of these material increase with increase in temperature.
10. Examples : Polyesters, epoxides, phenolics, phenol formaldehyde etc.	10. Examples : Polyethylene polystyrene, vinyl resin, polyamide.

10.12. PROCESSING OF THERMOSETTING PLASTIC

There are number of methods which are used for processing of thermosetting plastics. Commonly used methods are :

10.12.1. Compression Moulding

Compression moulding is similar to closed die forging in which plastic in form of solid granules are placed into heated cavity. Punch descends downwards and applies pressure to compress the compound into required shape. To allow chemical changes mould is kept closed for sufficient time and polymerization takes place. When the product gets hardened, mould is opened and part is removed. It is economical for small products.

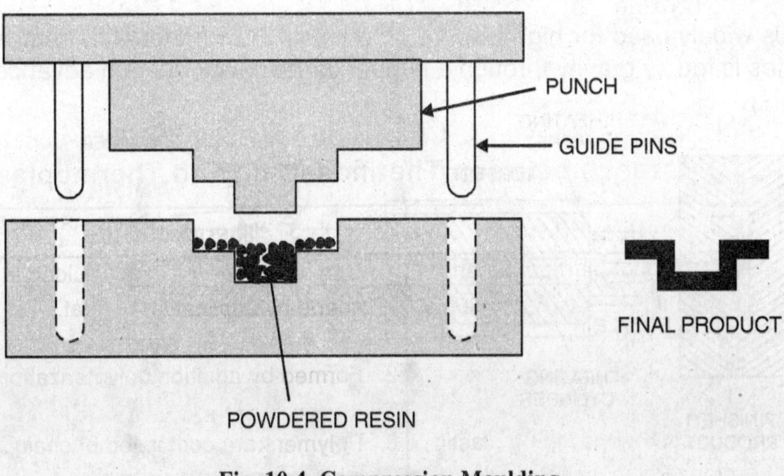

Fig. 10.4. Compression Moulding.

10.12.2. Transfer Moulding

It is the modification of compression moulding in which turbulence and uneven flow of molten plastic has been eliminated. Unpolymerized raw material is placed in plunger cavity where it is heated to molten state. Plunger move downwards and forces the molten metal to flow through sprues and gate into die cavities. Some advantages of this process are :

(i) Little pressure exists until the cavity is completely filled.

(ii) Operation is easy

(iii) It can be used where inserts are incorporated in the product

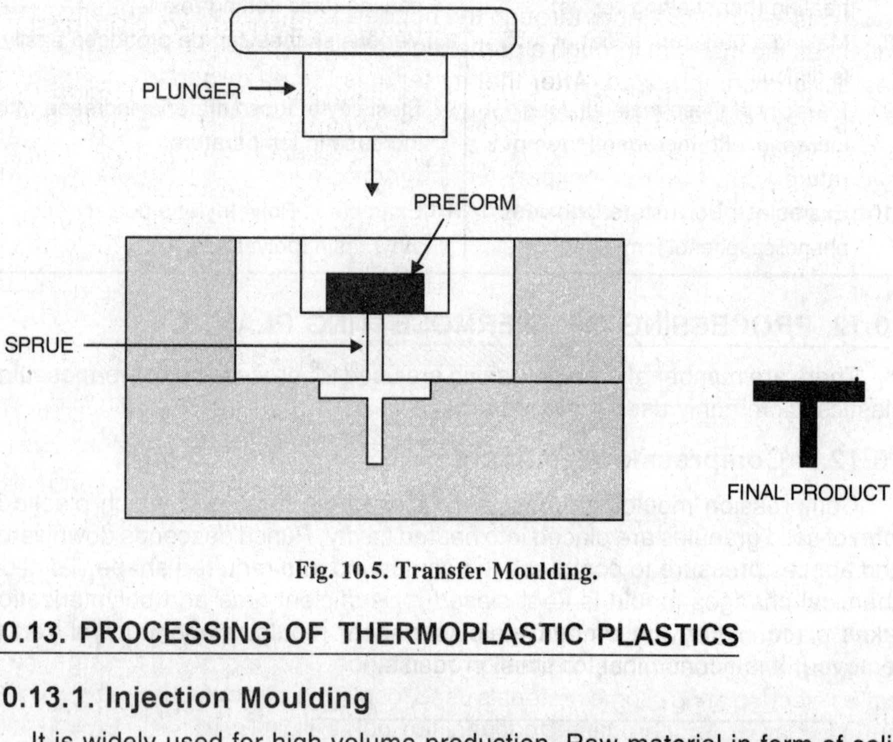

Fig. 10.5. Transfer Moulding.

10.13. PROCESSING OF THERMOPLASTIC PLASTICS

10.13.1. Injection Moulding

It is widely used for high volume production. Raw material in form of solid granules is fed by gravity through a hopper in the cavity. As ram advances, the

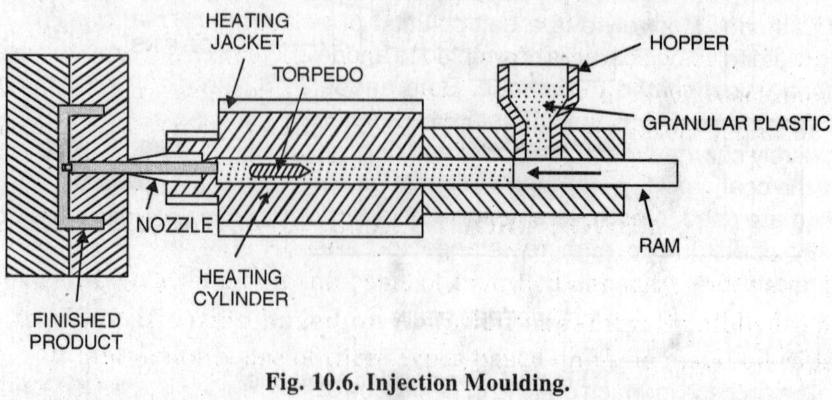

Fig. 10.6. Injection Moulding.

material is forced forward in a heated chamber. After that material is forced through thin section, called torpedo section where it is heated, melted and superheated in range of 200°C to 300°C. The material being under compression softens and is forced into the mould cavity through the nozzle where material solidifies on cooling and product is formed. It is fast process and is suitable for making parts with complex threads and inticrate shapes. It is commonly used for thermoplastic plastics.

10.13.2. Extrusion Moulding

This process is suitable for thermoplastic material. Material can be extruded into various shapes like sheets, ropes, pipes, etc. Thermoplastic powder is fed into chamber through the hopper. In this chamber rotating screw advances the material through a preheating section where material is uniformly heated, and compressed. After that material is forced through heated orifice of the die. Shape of product depends upon the size of orifice. Product is cooled by jets of air and sprays of water to harden it sufficiently and is carried away by conveyor. It is inexpensive and continuous method and is used to form tubes, pipes and even coated wire.

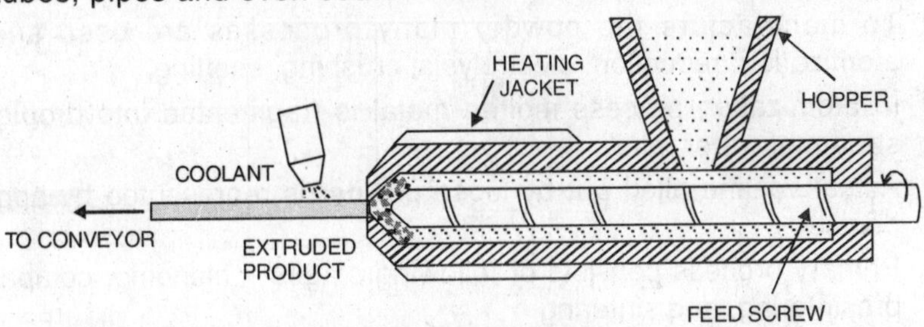

Fig. 10.7. Extrusion Moulding.

10.14. ELECTROPLATING [UPTU DEC. 2008]

The process in which metal and plastics are coated by means of electrolysis to impart wear resistance, corrosion resistance and improved appearance is called electroplating. Some materials used for plating of metallic and non-metallic surfaces are zinc, chromium, copper, tin, gold. Nickel plating provides good corrosion resistance but does not provide bright appearance.

This system consists of electrolyte, tank, low voltage D.C. supply, anode and cathode. Electrolyte is a salt solution of plating metal and is filled in the tank. Material to be coated is connected to anode and workpiece on which coating is done is connected to cathode. Both anode and cathode are immersed in electrolyte. When D.C voltage is applied metal ions move from anode towards negatively charged metal (cathode) and deposits on the surface of workpiece to form the coating. It is performed as continuous process. Gold, silver and platinum plating are used in both jewelry and electronics industries. Chromium plating is mostly used due to high resistance to corrosion and wear, and pleasing appearance.

10.15. GALVANIZING [UPTU DEC. 2008]

The process in which coating of highly corrosion resistant material is provided on the surface of metal by means of electro-chemical action is called galvanizing.

There are number of methods, which are employed for galvanizing in which hot dip galvanizing is widely used.

First step in galvanization process is to clean the sheet surface by removing rust, scale, oil, etc. Sheet is immersed into flux solution of zinc ammonium chloride and dried, this dried flux protects the sheet from oxidation during coating. The sheet is then annealed in a continuous furnace and immersed in a bath of molten zinc. Zinc and iron reacts metallurgically and produce a coating, which consist of zinc iron compound and surface layer of nearly pure zinc. After this, sheet are drawn through set of rollers to dry away the excess zinc and allows only the required amount of zinc to remain on the surface, which is sufficient to provide coating of desired thickness. It is used for coating on structural parts, wash tubs, telegraph wire, ships, tanks, transformer parts, etc.

SUMMARY

- Powder metallurgy is an art of manufacturing commercial articles from powdered metals and alloys.
- To manufacture the powder many processes are used such as atomization, reduction, electrolysis, crushing, shotting.
- In atomization process molten metal is fragmented into droplets by stream of water, air.
- Actual volume filled out by loose powder is represented by apparent density.
- Primary process consists of following stages – blending, compacting, presintering, and sintering.
- Blending is done to obtain uniform distribution of particle sizes. Blending is required when different metal powders are used or nonmetallic particles are added to impart some specific properties.
- In presintering, green compact is heated to a temperature below sintering temperature to increase the strength.
- In the sintering operation, compacted parts are heated to high temperature in controlled atmosphere but below the melting point of material being sintered.
- Secondary operation are performed to enhance precision, improve properties and special characteristics such as sizing, coining, machining, plating.
- Plastics are synthetic organic material, which can be formed into different shapes with the application of heat and pressure.

SHORT QUESTIONS WITH ANSWERS

Q.1. What is the use of powder metallurgy process?

Ans. It is used for mass production of small, intricate parts of high precision and eliminates the need for additional machining or finishing.

Q.2. Which reducing agents are used in reduction process of powder metallurgy?

Ans. Carbon Monoxide, hydrogen.

Q. 3. What is role of apparent density in powder metallurgy?

Ans. Portion of component can be made hard and dense by controlling apparent density.

Q. 4. What is the compression ratio for powder used in powder metallurgy?

Ans. Compression ratio is 2:1 to 3:1 for powder used in powder metallurgy and 10:1 for very fine powder.

Q. 5. What is the use of blending?

Ans. Blending is done to obtain uniform distribution of particle sizes. Blending is required when different metal powders are used or nonmetallic particles are added to impart some specific properties.

Q. 6. Why the two punches are used in compacting?

Ans. To obtain uniform distribution of pressure throughout the mass two punches are used.

Q. 7. What is use of sintering?

Ans. Sintering is done to increase the strength and hardness of the part to the require extent.

Q. 8. Which gases are required to make protective atmosphere during sintering?

Ans. Hydrogen, dissociated ammonia, or cracked hydrocarbons.

Q. 9. Why the secondary process are required after primary process?

Ans. Secondary operations are performed to enhance precision, improve properties and special characteristics.

Q. 10. What is the requirement of plating of sintered part?

Ans. Plating of sintered part is done to prevent it from corrosion and provide protective atmosphere.

Q. 11. Name the two major advantages in favour of powder metallurgy process. **[UPTU Dec. 2009]**

Ans. Production rate is high and labour cost is low.

Q. 12. In which process coating of highly corrosion resistant material is done by means of electrochemical reaction. **[UPTU May 2010]**

Ans. Galvanizing

REVIEW QUESTIONS

1. Explain the term powder metallurgy. Enlist stages of powder metallurgy process.
2. What are different processes used in production of metal powder?
3. Explain the different properties of metal powder suitable for powder metallurgy process.
4. What are different primary processes?
5. Explain secondary processes.
6. Explain advantages, disadvantages and applications of powder metallurgy.
7. Define plastics. Give the classification of plastics.
8. How thermosetting plastic differs from thermoplastic plastics?
9. Explain different properties of plastics.
10. Explain the following with neat sketches:
    (a) Compression Moulding (b) Transfer Moulding
    (c) Injection Moulding (d) Extrusion Moulding

11. Explain briefly the following with suitable examples.

(a) Electroplating (b) Galvanizing.

[UPTU Dec. 2008; May 2010]

12. Write short notes on presintering. **[UPTU Dec. 2011]**

SELF TEST QUESTIONS

A. State Whether True or False

1. The product obtained from powder metallurgy process requires high machining.
2. Carbon monoxide is used as reducing agent in reduction process of powder metallurgy.
3. Irregular shaped particles of powder provide poor green strength.
4. Coarse grain structure decreases the density of metal.
5. Presintering is applied where machining is not required.
6. Secondary operations enhance precision and improve characteristics.
7. Plating of sintered part is done to prevent the corrosion.
8. Compression moulding is suitable for processing of thermoplastic plastics.
9. Injection moulding is suitable for large volume production.
10. Powder metallurgy process provides good surface finish.

B. Fill in The Blanks

1. In ....process molten metal is fragmented into droplets by stream of water, air.
2. .... is defined as actual volume filled out by loose powder.
3. ....is done to obtain uniform distribution of particle sizes.
4. In presintering operation green compact is heated to temperature .... sintering temperature.
5. ....provides self-lubricating properties to sintered parts.
6. Plastics are .... resistance to corrosion and have.......... melting point.
7. .... plastics can be repeatedly softened by heating and hardened by cooling.
8. In hot dipping method coating of .... is provided on metallic surface.
9. .... is the modification of compression moulding.
10. .... moulding is used to form sheets, ropes, pipe, etc.

C. Multiple Choice Questions

1. Which one is not the stage of powder metallurgy?

   (a) Compacting (b) Tempering

   (c) Sintering (d) Blending
2. Apparent density depends upon

   (a) Particle size (b) Particle shape

   (c) Distribution of particles (d) All of the above.
3. Which operation is required to close the voids and provide the desired density?

   (a) Sizing (b) Plating

   (c) Coining (d) Impregnation

4. Which operation is used to increase the porosity in sintered compact?
(a) Sizing (b) Plating
(c) Infiltration (d) Impregnation

5. Which moulding is preferred for processing of thermoplastic plastics?
(a) Injection (b) Compression
(c) Transfer (d) None.

6. In which process coating of highly corrosion resistant material is done by means of electrochemical reaction.
(a) Normalizing (b) Austempering
(c) Galvanizing (d) None

7. Which process is the primary process of powder metallurgy ?
(a) Sintering (b) Coining
(c) Sizing (d) Machining

8. Refractory metals are sintered upto .... percentage of their melting point.
(a) 60 (b) 80
(c) 70 (d) 90

9. For very fine powder compression ratio is
(a) 2:1 (b) 8:1
(c) 4:1 (d) 10:1

10. Which one is stage of sintering operation ?
(a) Burn Off Chamber Stage (b) High temperature stage
(c) Cooling period stage (d) All of the above.

11. Molten metal is fragmented into droplets by using stream of water and air
(a) Blending (b) Atomization
(c) Compacting (d) Galvanizing

12. Size and shape of atomized particle depends upon
(a) Velocity of atomizing jets (b) Medium of atomizing jets
(c) Temperature of liquid (d) All of these

13. Reducing agent used in reduction process of powder manufacture is
(a) Oxygen (b) Hydrogen
(c) Nitrogen (d) H_2O

14. Process in which molten metal is passed through a sieve into water to produce large spherical particles
(a) Electrolysis (b) Milling
(c) Shotting (d) Atomization

15. ______ shaped particles achieve good green strength
(a) Irregular (b) Angular
(c) Lamellar (d) Flag

16. Density of metal ________ due to fine grain particles
(a) Decrease (b) Increases
(c) Not affected (d) None of these

17. Which process is done to obtain uniform distribution of particle size
(a) Sintering (b) Compacting
(c) Presintering (d) Blending

18. Powders are compacted and densified into accurate size or shape, known as

(a) Green compact (b) Black compact
(c) Blue compact (d) Red compact

19. In presintering process, green compact is heated to temperature ______ sintering temperature to increase the strength

(a) Above (b) Below
(c) Equal (d) None of these

20. Process in which compacted parts are heated to high temperature in controlled atmosphere, but below the melting point of material being sintered

(a) Presintering (b) Sintering
(c) Blending (d) Briquetting

21. Process in which pores in sintered part are filled with oil and provide self lubricating properties to parts is

(a) Sizing (b) Coining
(c) Infiltration (d) Impregnation

22. Which method is used for processing of thermosetting plastics

(a) Injection moulding (b) Extrusion moulding
(c) Compression moulding (d) None of these

23. Which method is the modification of compression moulding in which turbulence and uneven flow of molten plastic has been eliminated

(a) Injection moulding (b) Transfer moulding
(c) Extrusion moulding (d) None of these

24. Process in which metal and plastics are coated by means of electrolysis.

(a) Galvanizing (b) Moulding
(c) Electroplating (d) Machining

25. Process in which coating of highly corrosion resistant material is produced on the surface of metal by means of electrochemical action

(a) Galvanizing (b) Electroplating
(c) Electrolysis (d) Extrusion

D. Match the Following

1.

(A)	(B)
(i) Blending	(a) Primary process of powder metallurgy
(ii) Thermosetting plastic	(b) Injection moulding
(iii) Thermoplastic	(c) Secondary operation of powder metallurgy
(iv) Coining	(d) Compression moulding

2.

(A)	(B)
(i) Sizing	(a) Close the voids and provide desired density
(ii) Coining	(b) provide self lubricating property
(iii) Infiltration	(c) Correct the dimension of sintered part
(iv) Impregnation	(d) Fill the porosity in sintered compact

3.

(A)	(B)
(i) Car body	(a) Machining
(ii) Clutch lining	(b) Casting
(iii) Gears	(c) Sheet metal working
(iv) Engine block	(d) Powder metallurgy

4. (A) (B)

(i) Mild steel angle and channel (a) Welding
(ii) Carburettors (b) Forging
(iii) Roof stress (c) Casting
(iv) Gear wheels (d) Rolling

ANSWERS

A. True or False

1. False 2. True 3. False 4. True 5. False
6. True 7. True 8. False 9. True 10. True

B. Fill in the Blanks

1. Atomization
2. Apparent density
3. Blending
4. Below
5. Impregnation
6. High, Low
7. Thermoplastic
8. Zinc
9. Transfer moulding
10. Extrusion

C. Multiple Choice Questions

1. (b) 2. (d) 3. (c) 4. (c) 5. (a)
6. (c) 7. (a) 8. (d) 9. (d) 10. (d)
11. (b) 12. (d) 13. (b) 14. (c) 15. (a)
16. (b) 17. (d) 18. (a) 19. (b) 20. (b)
21. (d) 22. (c) 23. (b) 24. (c) 25. (a)

D. Match the Following

1. (i)—(a), (ii)—(d), (iii)—(b), (iv)—(c) 2. (i)—(c), (ii)—(a), (iii)—(d), (iv)—(b)
3. (i)—(c), (ii)—(d), (iii)—(a), (iv)—(b) 4. (i)—(d), (ii)—(c), (iii)—(a), (iv)—(b)

❑❑❑

MODEL QUESTION PAPERS

MODEL QUESTION PAPER–1

Time : 2 Hours *Total Marks : 50*

Section-A

Attempt all the questions :

1. This section contains 10 questions. (Objective/Fill in the blanks types choose/Fill the correct answer.

(a) Formation of new grains, due to plastic deformation is called ______.

(b) ______ plastic can be repeatedly softened by heating and hardened by cooling.

(c) Gas welding is ______ process

(i) Forge welding (ii) Fusion welding
(iii) Pressure welding (iv) None

(d) Pouring basin is used to avoid ______

(i) Oxidation of metal (ii) Rapid cooling
(iii) Mould erosion (iv) None

(e) In this operation large numbers of closely spaced holes are produced

(i) Shaving (ii) Perforating
(iii) Trimming (iv) Lancing

(f) Which metal is used in storage batteries ?

(i) Lead (ii) Tin
(iii) Zinc (iv) Steel

(g) Process of adding nitrogen and carbon to surface is

(i) Cyaniding (ii) Nitriding
(iii) Quenching (iv) All of the above

(h) Surface grinders are used to produce

(i) Flat surfaces (ii) Circular surface
(iii) Irregular surface (iv) None

(i) Brittleness shows the lack of

(i) Elasticity (ii) Plasticity
(iii) Resilience (iv) Toughness

(j) Which process is the primary process of powder metallurgy ?

(i) Sintering (ii) Coining

(iii) Sizing (iv) Machining

Section-B

2. Attempt any **three** questions. All questions carry equal marks

(a) Explain the following types of cast iron with their composition, properties and uses :

(i) Gray cast iron

(ii) White cast iron

(iii) Nodular cast iron

(iv) Mottled cast iron

(b) What do you mean by heat treatment ? Explain the objectives of heat treatment and its prooess ?

(c) Explain the rolling operation with neat sketch and explain different types of rolling defects.

(d) Explain the resistance spot welding with neat sketch and mention the advantages and its applications.

(e) Explain plastics with their properties. Differentiate between thermosetting plastic and thermoplastic.

Section-C

Attempt all questions.

3. Answer any one part of the following :

(a) With the help of sketch explain the creep curve.

(b) What are the important properties of aluminium ? Describe the two important aluminium alloys.

4. Answer any one part of the following :

(a) With the help of neat sketch explain the different components of die and punch assembly.

(b) Explain the function and operation of cupola furnace with neat sketch.

5. Answer any one part of the following :

(a) Describe the basic working principle and important parts of planer with sketch. Also describe common operations performed on planer.

(b) What is oxyacctylene welding. Explain different types of equipment used in oxyaccetylene welding with neat sketches.

6. (a) Explain different types of plant layout with their advantage and disadvantages.

(b) What do you mean by powder metallurgy. What are different processes used in production of metal powder.

MODEL QUESTION PAPER–2

Time : 2 Hours *Total Marks : 50*

Section-A

Attempt all the questions :

1. _______ is a property which resists deflection or deformation.

2. In white cast iron, carbon is present in form of

(a) Flakes of graphite (b) Carbide

(c) Nodular aggregates (d) Spheroids

3. Martensite is the hardest constituent of iron carbon system. (T/F)

4. _______ defect is due to misalignment of die halves

(a) Cold shut (b) Flakes

(c) Die shift (d) Scale pits

5. _______ is also called flashless forging.

6. Good temperature gradient is achieved in

(a) Top gate (b) Bottom gate

(c) Parting gate (d) None

7. Cross slide is used to provide transverse tool movement. (True/False)

8. To produce a tapered surface on lathe, tool must move

(a) At an inclination with workpiece axis

(b) Parallel to workpiece axis

(c) Perpendicular to workpiece axis

(d) All of the above.

9. Which metal is welded by using oxidizing flame ?

(a) Brass (b) Aluminium

(c) Lead (d) Stainless steel

10. Match the following

(A)	(B)
(i) Thermosetting plastic	(a) Polyethylene
(ii) Thermoplastic	(b) Polyurethane
(iii) Natural rubber	(b) Polyester
(iv) Synthetic rubber	(b) Latex

Section-B

Attempt any three questions :

1. Explain the following :

(a) Brittle fracture (b) Ductile fracture

(c) Fatigue fracture (d) Creep fracture

2. What are the properties and uses of aluminium alloy ?

3. Explain wire drawing with neat sketch.

4. Explain the procedure of making mould ?

5. Define plant location. Explain various factors that are considered for plant location.

Section-C

Attempt any five questions :

1. Explain various methods of taper turning.
2. Explain different types of flames used in oxy-acetylene welding.
3. What is tempering ? Explain the different stages of tempering.
4. Explain different properties of metal powder suitable for powder metallurgy process.
5. Explain different components of gating system used in casting.
6. What do you mean by sheet metal operation. Explain various sheet metal operations.

MODEL QUESTION PAPER–3

Time : 2 Hours *Total Marks : 50*

Section-A

Attempt all the questions :

1. Primary creep occurs at ______ temperature.
2. Carbide present in white cast iron is known as ______.
3. Process of adding nitrogen and carbon to surface is
   (a) Cyaniding (b) Nitriding
   (c) Quenching (d) All of the above
4. In ______ processes parts are produced without formation of chip.
5. Which forging is used to form bolts
   (a) Upset forging (b) Press forging
   (c) Roll forging (d) Swaging
6. Dross are light impurities at top surface of casting. (True/False)
7. Shaper is used for machining of large flat surface. (True/False)
8. Match the following

(A)	(B)
(i) Lead screw	(a) Lathe
(ii) Rocker arm	(b) Milling machine
(iii) Universal indexing	(b) Shaper
(iv) Flute	(b) Drilling machine

9. Time period during which current flows to raise the temperature is
   (a) Squeeze time (b) Hold time
   (c) Weld time (d) Off time
10. Wood is a composite of ______ and ______.

Section-B

Attempt any three questions :

1. What is high speed steel ? Explain its type with examples.
2. How the wrought iron is obtained. Explain different properties and applications.
3. Explain different type of forging defects.
4. Explain different types of moulding sand used in casting.
5. Compare A.C. and D.C. welding.

Section-C

Attempt any five questions :

1. Explain different type of composite materials with neat sketch.
2. Explain the resistance projection welding with neat sketch. Also mention advantages, disadvantages and applications.
3. Discuss working principle of milling operation. Explain difference between up milling and down milling.
4. What is the use of core in casting. Classify different types of core.
5. Explain rolling operation. Classify different types of rolling stand.
6. What is normalising ? Why is it done ? Also mention the difference between normalizing and annealing.

MODEL QUESTION PAPER–4

Time : 2 Hours *Total Marks : 50*

Section-A

Attempt all the questions :

1. Percentage of carbon in eutectoid steel is ______.
2. Match the following

(A)	(B)
(i) Ferrite	(a) FCC structure
(ii) Austenite	(b) BCC structure
(iii) Cementite	(c) Needle like structure
(iv) Martensite	(d) Carbide of iron

3. ______ plate is used for locating and supporting the die and punch assembly.
4. Which part is used in extrusion process to form hollow tubes
   (a) Ram (b) Plunger
   (c) Mandrel (d) None
5. Blow holes are produced by ______.

6. Upper part of moulding flask is known as
   (a) Cope (b) Drag
   (c) Chaplet (d) Pattern
7. In shaper
   (a) Workpiece reciprocates past a stationary tool
   (b) Tool reciprocates past a stationary workpiece
   (c) Tool and workpiece both reciprocates
   (d) Tool rotates with respect to workpiece
8. In ______ welding polarity is not fixed at terminal.
9. Hard solder is an alloy of
   (a) Lead and tin (b) Copper and zinc
   (c) Copper and lead (d) Zinc and tin
10. Product layout is used for
   (a) Continuous production (b) Batch production
   (c) Job shop production (d) All of the above

Section-B

Attempt any three questions :

1. What is stainless steel ? Explain different types of stainless steel with examples.
2. Explain different type of rolling defects.
3. What is casting ? Explain the casting terminology with neat sketch.
4. Explain different types of joints used in welding with neat sketch.
5. Define production and productivity. Also explain the difference between them.

Section-C

Attempt any five questions :

1. Explain the following with neat sketches
   (a) Compression moulding
   (b) Transfer moulding
2. List various properties of rubber. Also classify rubber with examples.
3. Define soldering. Explain procedure of soldering and its advantages.
4. Define grinding. Explain different type of cylindrical grinder with neat sketches.
5. Explain different properties of moulding sand.
6. Explain following process
   (a) Nitriding
   (b) Induction hardening
   (c) Cyaniding

❑❑❑

PREVIOUS YEARS EXAMINATION PAPERS

B.TECH

Ist Semester (2008-09)

MANUFACTURING PROCESSES

Time : 2 Hours *Total Marks : 50*

Note : Be precise and scientific in writing.

Section-A

Attempt all the questions :

1. This section contains 10 questions. (Objective/Fill in the blanks types). Choose/Fill the correct answer.

(a) Manufacturing is the process of coverting raw material into ______

(b) The ductility of a piece of chalk is ______

(c) The % of carbon in medium carbon steel varies

(i) Less than 0.20 (ii) 0.10 to 0.30

(iii) 0.30 to 0.60 (iv) More than 0.60.

(d) Stainless steel comes under the category of

(i) Non ferrous metal and alloy

(ii) Ferrous metal and alloy

(iii) Polymer (iv) Composite

(e) Bronze is an alloy of

(i) Aluminium and zinc (ii) Aluminium and nickel

(iii) Copper and tin (iv) Copper and zinc

(f) Forging of metal is a

(i) Machining process (ii) Welding process

(iii) Joining process (iv) Metal forming process

(g) Blow holes are the common defects found in

(i) Welding (ii) Machining

(iii) Forming (iv) Casting

(h) Finishing is the operation best done by

(i) Milling machine (ii) Shaper machine

(iii) Planer machine (iv) Grinding

(i) Swing is the one of specifications of

(i) Lathe machine (ii) Shaper machine

(iii) Drilling machine (iv) Grinding machine

Section-B

2. Attempt any **three** questions. All questions carry equal marks :

(a) Explain the following terms :

(i) Ductility (ii) Stiffness (iii) Brittleness

(iv) Fatigue (v) Malleability

(b) Draw rough sketch of the stress strain diagrams of

(i) Ductile material and

(ii) Brittle material

Briefly explain, what do you understand by Toughness.

(c) Explain

(i) Mould and (ii) Pattern.

Describe the two important pattern allowances.

(d) With the help of sketch, explain the basic components of lathe machine and various operations performed on it.

(e) With the help of sketch, define the following operations

(i) Forging (ii) Rolling

(iii) Drawing (iv) Extrusion

Section-C

Note : Attempt any **four** questions. All questions carry equal marks.

3. Answer any one part of the following :

(a) Based on % of carbon, classify carbon steels. How do stainless steel become stainless ?

(b) What are the major uses of copper ? What are the alloying elements in brass and bronze, respectively and also write its uses.

4. Answer any **one** part of the following :

(a) Differentiate between hot and cold working of metals. Bring out the advantages and disadvantages of each of these techniques. Explain with neat sketch the rolling of steel bar.

(b) Define the following terms with sketch as used in sand casting :

(i) Core (ii) Core-Prints (iii) Sprue

(iv) Runner (v) Riser

5. Answer any **one** part of the following :

(a) With the help of schematic sketch, describe the basic working principle and important parts of shaper machine. Also describe the difference in operation between shaper and planer.

(b) Explain fusion as it relates to welding operations. How will you classify the welding processes ? Explain the features of neutral, reducing and oxidizing flames.

6. Answer any **one** part of the following :

(a) Discuss the role and importance of materials and manufacturing for the growth of any nation. Explain the production and productivity.

(b) Explain briefly the following with suitable examples :

(i) Electroplating (ii) Galvanizing

(iii) Plant layout (iv) Ceramics

B.TECH

IInd Semester (2008-09)

MANUFACTURING PROCESSES

Time : 2 Hours *Total Marks : 50*

Note : Be precise and scientific in writing.

Section-A

Attempt all the questions : **10 × 1 = 10**

1. This section contains 10 questions. (True/False, Objective/Fill in the blanks types). Choose/Fill the correct answer.

(1) The castability of Al is very poor. (True/False)

(2) The machinability of Nickel and Steel is good. (True/False)

(3) Hot working is carried out below the recrystallization temperature. (True/False)

(4) Name the **two** defects in metal formed product.

(5) Fatigue failure occurs when a part is subjected to

(i) Compressive stress (ii) Tensile stress

(iii) Fluctuating stress (iv) Uniform stress.

(6) Press working of metal is a

(i) Machining process (ii) Welding process

(iii) Joining process (iv) Metal forming process

(7) Core is used to produce ________ in casting.

(8) Allowance provided on surfaces parallel to direction of withdrawn in casting is known ________.

(9) In compound rest method of taper turning, compound rest is swilled by the angle :

(i) $\tan \alpha = \frac{d_1 - d_2}{l}$ (ii) $\tan \alpha = \frac{d_1 - d_2}{2l}$

(iii) $\tan \alpha = L \times \frac{d_1 - d_2}{l}$ (iv) $\tan \alpha = \frac{d_1 - d_2}{2}$

(10) Match the following

(A)	(B)
(i) Hard wood	(a) Glass
(ii) Ceramic	(b) Deodar
(iii) Soft wood	(c) Shisham
(iv) Composite material	(d) Tufnol

Section-B

2. Attempt any **three** questions. All questions carry equal marks : **3 × 5 = 15**

(a) Explain the following terms :

(i) Strength (ii) Toughness

(iii) Fatigue (iv) Malleability

(b) Classify the carbon steel based on the percentage of carbon as low, mild, medium and high carbon steel. Write their properties and applications.

(c) Explain with neat sketch, the basic working principle of extrusion. Describe its applications in industry. Name the four extruded products.

(d) With help of neat sketch, explain the basic components of shaper machine and various operations performed on it.

(e) With the help of neat sketch, define the following operations :

(i) Forging (ii) Rolling

(iii) Taper turning (iv) Galvanizing.

Section-C

Note : Attempt any **three** questions. Each question carries equal marks.

5 × 5 = 25

3. Why heat treatment of carbon steel is required ? Give the elementary idea of various annealing operations.

4. Differentiate between hot and cold working of metals. Classify the metal forming operations and mention their specific applications.

5. Why the pattern is different from casting ? Describe with neat sketch, the design steps in casting process. What is the function of core in casting ?

6. With the help of schematic sketch, describe the basic working principle of milling machine. Differentiate between up milling and down milling.

7. How will you classify the welding processes ? Explain with suitable applications, the working principle of resistance welding.

8. Discuss the role and importance of materials and manufacturing for the growth of any nation. Explain with suitable examples, the types of production.

B.TECH
(Semester I) ODD Semester Theory Examination (2009-10)
MANUFACTURING PROCESSES

Time : 2 Hours *Total Marks : 50*

Note : Be precise and scientific in writing.

Section-A

*Attempt **all** the questions.*

This section contains **10** questions. **All** questions carry **equal** marks. **10**

1. Energy required to rupture the material is known as ....
2. List the alloying element of duralumin.
3. Misrun is a metal forming defect. (True/False)
4. Write the two welding defects.
5. Brittle failure occurs when a part is subjected to :
   (i) Compressive stress (ii) Tensile stress
   (iii) Fluctuating stress (iv) Uniform stress.
6. Finishing is :
   (i) Machining process (ii) Welding process
   (iii) Joining process (iv) Metal forming process
7. Hole is produced in casting with the help of ....
8. Molding sand is called green sand because it contains :
   (i) Bakelite (ii) Moisture
   (iii) No moisture (iv) Dry sand
9. In compound rest method of taper turning, compound rest is swilled by the angle :
   (i) $\tan\alpha = \frac{d_1 -, d_2}{l}$ (ii) $\tan\alpha = \frac{d_1 - d_2}{2l}$
   (iii) $\tan\alpha = L \times \frac{d_1 - d_2}{l}$ (iv) $\tan\alpha = \frac{d_1 - d_2}{4l}$
10. Name the two major advantages in favour of powder metallurgy process.

Section-B

Attempt any **three** questions. All questions carry equal marks : **3 × 5 = 15**

1. Explain the following terms with suitable example :
   (i) Ductility (ii) Toughness
   (iii) Creep

2. Explain the properties and applications of :
   (i) Wrought Iron (ii) Tool Steel
   (iii) Cast Iron
3. Explain with near sketch, the basic working principle of Rolling. Describe its application in industry.
4. Differentiate between shaper and planner. With the help of neat sketch, explain the basic components of lathe machine.
5. Describe with suitable examples, plant layout and its different types and applications.

Section-C

Note : Attempt any **three** questions.
Each question carries equal marks. **5 × 5 = 25**

1. Write short notes on Annealing and Normalizing of carbon steels.
2. Differentiate between hot and cold working of metals. Bring out the advantages and disadvantages of each of these techniques.
3. With the help of neat diagram, explain the working principle of cupola. Also write its limitations.
4. Distinguish between welding, brazing and soldering processes. Write about the importance of fluxes being used in welding.
5. With the help of schematic sketch, describe the basic working principle of grinding process.
6. Write short notes on :
   (i) Ceramics and its applications
   (ii) Composite materials and their applications

B.TECH

IInd Semester Even Semester Theory Examination, 2009-10

MANUFACTURING PROCESSES

Time : 2 Hours *Total Marks : 50*

Note : *Attempt all questions. Be precise and scientific in writting.*

Section-A

1. This section contains 10 questions. All questions are compulsory : **10 × 1 = 10**

(a) The castability of cast iron is poor. (True/False).

(b) Stainless steel is an alloy steel. (True/False).

(c) Define malleability.

(d) Bronze is an alloy of :

(i) Aluminium and zinc (ii) Aluminium and nickel
(iii) Copper and tin (iv) Copper and zinc

(e) Write two casting defects.

(f) Ductile failure occurs when a part is subjected to :

(i) Compressive Stress (ii) Tensile Stress
(iii) Fluctuating Stress (iv) Uniform Stress

(g) Material removal is a :

(i) Machining process (ii) Welding process
(iii) Joining process (iv) Metal forming process

(h) Hole is produced in casting with the help of ....

(i) Write two properties of molding sand.

(j) Name the two major advantages in favour of powder metallurgy process.

Section-B

2. Attempt any **three** questions. All questions carry equal marks : **3 × 5 = 15**

(a) Explain the following terms :

(i) Fatigue (ii) Stiffness
(iii) Creep (iv) Brittle fracture
(v) Elasticity

(b) Explain with neat sketch and suitable example, the stress-strain diagram of :

(i) Ductile material and (ii) Brittle material

(c) With the help of neat sketch, explain the basic components of lathe machine and various operations performed on it.

(d) Write short notes on :

(i) Galvanizing (ii) Electroplating

(e) Describe with suitable example, plant layout and its different types and applications.

Section-C

Question (a) is **compulsory** and carries **Seven** marks. Attempt **any three** more questions from the remaining (all **carry equal** marks of **Six**). **(25)**

(a) Name three alloys each of ferrous and non-ferrous metals. Write their applications.

(b) Differentiate between hot and cold working of metals. Mention two advantages and disadvantages of each of these techniques.

(c) Why the pattern is different from casting ? Describe with neat sketch, the various steps in casting process. What is the function of core in casting ?

(d) How will you classify the welding processes ? Explain with suitable applications, the working principle of resistance welding.

(e) Discuss the role and importance of Materials and Manufacturing for the growth of any nation. Explain with suitable examples, the types of production.

(f) Write short notes on :

(i) Ceramics and their applications.

(ii) Composite materials and their applications.

B.TECH

First Semester ODD Semester Theory Examination, 2010-11

MANUFACTURING PROCESSES

Time : 2 Hours *Total Marks : 50*

Section-A

1. Attempt **all** questions. All questions carry equal marks. **10 × 1 = 10**

(A) Non-metallic elements which are basically the oxides and compounds of metal and non-metals are known as :

(a) Composite (b) Organic Polymers
(c) Ceramics (d) Ores

(B) Pudding furnace is lined inside with ....

(C) Which coolant has highest cooling rate in heat treatment process ?

(a) Water (b) Sulphuric Acid
(c) Air (d) Brine

(D) The sheet metal operation in which metal is removed in small increments :

(a) Perforating (b) Punching
(c) Lancing (d) Nibbling

(E) In casting process, inclusions are :

(a) Steel particles (b) Metallic particles
(c) Iron particles (d) Non-Metallic particles

(F) Reaming operation is performed to :

(a) Enlarge a previously drilled hole
(b) Finish previously drilled hole to accurate size
(c) Both (a) and (b)
(d) None of the above

(G) In resistance welding the time period during which current flows to rise the temperature is :

(a) weld time (b) on time
(c) off time (d) squeeze time

(H) In carburizing flame the maximum temperature produced is :

(a) about 3200°C (b) about 3500°C
(c) about 3000°C (d) None of the above

(I) Inorganic non-metallic material which are used at very high temperature :

(a) Cement (b) Rubber
(c) Thermosetting plastic (d) Ceramic

(J) Combination layour combines the features of :

(a) Product and process layout
(b) Jobshop and process layout
(c) Jobshop and product layout
(d) Fixed position and process layout

Section-B

2. Attempt any **three** questions. All questions carry equal marks. **3 × 5 = 15**

(A) Explain the following :

(a) Ductile fracture (b) Brittle fracture
(c) Creep fracture (d) Fatigue fracture

(B) Explain different properties of moulding sand.

(C) What is the principle of gas welding ? Explain different types of oxyacetylene flames.

(D) What are the objectives of plant layout ? Explain different types of layout with their advantages and disadvantages.

(E) Explain the working principle and operation of a Millng machine with neat diagram.

Section-C

3. Attempt any **two** parts : **(2 × 5 = 10)**

(A) Classify steel on the basis of carbon percentage. Also write properties and uses of them.

(B) What do you mean by case-hardening ? Explain different method of case hardening in detail.

(C) Differentiate Cast Iron on the basis of percentage of carbon. Explain with neat diagram the construction and working of cupola furnace.

4. Attempt any **two** parts : **(2 × 7.5 = 15)**

(A) Define weldability. Explain in brief the electric arc welding and different equipment used in electric arc welding. Also explain its advantages and disadvantages.

(B) Write short notes on any three of the following :

(a) Soldering
(b) Brazing
(c) Centrifugal casting
(d) Die casting process.

(C) With the help of schematic diagram describe the basic working principle and important parts of drilling machine. Also describe drilling operations.

B.TECH
(Second Semester) Even Semester Theory Examination (2010-11)
MANUFACTURING PROCESSES

Time : 2 Hours *Total Marks : 50*

Note : Be precise and scientific in writing.

Section-A

*Attempt **all** the questions.*

This Section contains 10 questions (objective/fill in the blanks types).

Choose/Fill the correct answer.

(1 × 10 = 10)

1. The castability of Al is very poor. (True/False)

2. The machinability of Nickel and Steel is good. (True/False)

3. Energy required to rupture the material is known as ....

4. Brittle failure occurs when a part is subjected to :

(i) Compressive stress (ii) Tensile stress

(iii) Fluctuating stress (iv) Uniform stress

5. Stainless steel comes under the category of :

(i) Non-ferrous metal and alloy

(ii) Ferrous metal and alloy

(iii) Polymer

(iv) Composite.

6. Bronze is an alloy of :

(i) Aluminium and Zinc (ii) Aluminium and Nickel

(iii) Copper and Tin (iv) Copper and Zinc

7. Hole is produced in casting with the help of ....

8. Forging of metal is a :

(i) Machining Process (ii) Welding Process

(iii) Joining Process (iv) Metal Forming Process

9. Metal forming is the process in which :

(i) Removal of metal takes place

(ii) Shaping of metal by compressive forces applied through various dies and tools.

(iii) Some additional materials are joined to the raw material

(iv) Neither material removal non-addition.

10. Name the two major advantages in favour of powder metallurgy process.

Section-B

Attempt any **three** questions. All questions carry equal marks : **5 × 3 = 15**

1. Classify the carbon steel based on the percentage of carbon as low, mild, medium, and high carbon steel. Write their properties and applications.
2. Explain with neat sketch the stress-strain diagram of a
   (i) Ductile material
   (ii) Brittle material
3. Explain (i) mould and (ii) pattern. Describe the four important pattern allowances.
4. Differentiate between shaper and planner. With the help of neat sketch, explain the basic components of lathe machine.
5. With the help of neat sketch, define the following operations :
   (i) Forging (ii) Rolling
   (iii) Drawing (iv) Extrusion

Section-C

Note : Attempt any **five** questions. Each question carries equal marks.

(5 × 5 = 25)

1. What are the major uses of copper ? What are the alloying elements in brass and bronze, respectively ? List various types of brasses. Also give their application.
2. Differentiate between hot and cold working of metals. Bring out the advantages and disadvantages of each of these techniques. Explain with neat sketch the forging of ingots.
3. Define the following terms with neat sketch as used in sand casting :
   (i) Core (ii) Core prints
   (iii) Sprue (iv) Runner
   (v) Riser
4. With the help of schematic sketch, describe the basic working principle and important parts of shaper machine. Also describe the difference in operation between shaper and planer.
5. Explain fusion as it relates to welding operations. How will you classify the welding processes ? Explain the features of neutral, reducing and oxidizing flames. Also draw labeled diagrams of these flames.
6. Discuss the role and importance of materials and manufacturing for the growth of any nation. Explain production and productivity.
7. Explain the following with suitable examples :
   (i) Electroplating (ii) Galvanizing
   (iii) Plant layout (iv) Ceramics.

B.TECH.

(Semester-I) Theory Examination (2011-12)

MANUFACTURING PROCESSES

Time : 2 Hours *Total Marks : 50*

Note : (i) Attempt questions from each Section.

(ii) Be precise and scientific in writing.

Section-A

1. Attempt *all* parts : **(1 × 10 = 10)**

(a) Write the major difference between toughness and hardness of a material.

(b) What is the basic aim of tempering ?

(c) List four hot working processes.

(d) Define Gating ratio.

(e) List four important casting defects.

(f) How lathe size is specified ?

(g) What HAZ ?

(h) Write the composition and percentage of alloying elements present in Brass and bronze.

(i) Write two difference between Soldering and Brazing.

(j) Define Production and Productivity.

Section-B

1. Attempt any *three* parts : **5 × 3 = 15**

(a) Explain the following terms :

(i) Resilience

(ii) Fatigue

(iii) Strength

(iv) Stiffness

(v) Ductile Fracture.

(b) Differentiate between normalizing and annealing in terms of the process and nature of the product.

(c) List the advantages of forging of metals. Why is press forging preferred over hammer forging process.

(d) How does a shaper differ from a planar in terms of their operation and type of work pieces ?

(e) Write short notes on the following :

(i) Powder Metallurgy

(ii) Pre-sintering.

Section-C

Attempt *all* questions : **(5 × 5 = 25)**

3. What are high speed steels ? What are the principal alloying elements in them ? Describe their properties in brief.

Or

State the difference between steel and cast iron with respect to their composition, tensile strength, hardness and toughness.

4. Describe the various types of pattern. Write the pattern allowances. How do pattern differ from casting ?

Or

Draw a neat sketch of die-punch assembly with accessory parts or components. Explain how cutting operations such as blanking or piercing differ from deep drawing operations.

5. With the help of a suitable sketch, describe the geometry of a twist drill and also explain how drill sizes are designated.

Or

Draw the different types of flames used in gas welding. How would you identify these flames ? What are the specific uses of each of these flames ?

6. Mention advantages, disadvantages and applications of powder metallurgy. Describe briefly each step involved in powder metallurgy manufacturing processes.

Or

Explain briefly about the different types of plant layout. Write the various factors effecting plant layout.

7. What is a gating system ? Explain briefly about the different components of gating system.

Or

Define milling. What are the various work-holding devices used in milling ? Explain their relative applications.

B.TECH.

(Semester-II) Theory Examination, 2011-12

MANUFACTURING PROCESSES

Time : 2 Hours *Total Marks : 50*

Section-A

1. Attempt **all** questions. All questions carry equal marks : **10 × 1 = 10**

(A) Name two alloys of aluminum with compositions.

(B) Due to which type of loading the fatigue failure occurs ?

(C) Define the term recrystallization.

(D) What is the function of riser in casting process ?

(E) What is the name of product of cupola furnace ?

(F) What do you mean by gang milling ?

(G) Name the materials which are welded by neutral flame.

(H) Give the two examples of fixed position layout.

(I) In which process coating of highly corrosion resistant material is done by means of electrochemical reaction ?

(J) List the operations performed on Lathe machine.

Section-B

2. Attempt any **three** questions. All questions carry equal marks : **3 × 5 = 15**

(A) Compare the following :

(i) Hardness and toughness

(ii) Strength and stiffness

(B) Compare hot working and cold working processes with suitable examples. Also discuss the advantages and disadvantages of each.

(C) Differentiate between down milling and up milling. What are the various work holding devices used in milling ? Explain their relative applications and disadvantages.

(D) What are the differences between production and productivity ? Also discuss the product layout and process layout with suitable examples.

(E) Describe the type of flames used in gas welding. What are the applications of each ?

Section-C

3. Attempt any **two** parts : **(2 × 5 = 10)**

(A) (i) Differentiate between Ductile fracture and brittle fracture.

(ii) What do you understand by creep ? Discuss the creep curve.

(B) Describe the various kinds of patterns in use. What are the allowances provided, when making a pattern ? How does the pattern differ from casting required ?

(C) Explain the different properties of metal powder suitable for powder metallurgy process. Also discuss the applications of powder metallurgy.

4. Attempt any **two** parts : **(2 × 7½ = 15)**

(A) Write short notes on any **three** of the following :

(i) High speed steel

(ii) Moulding sand properties

(iii) Deep drawing process

(iv) Composite materials.

(B) What are the main difference between a shaper and planer ? Which are the drive mechanisms used in shaper ? Discuss any one in brief with neat sketch.

(C) How is an arc obtained in arc welding ? What are the different power sources used in welding ? What are the advantages and limitations of each ?

❑❑❑

Index

D

E

F

G

H

I

J

K

L

M

N

□□□

Win Prizes

ATTENTION READERS

Win Prizes

We request for your frank assessment regarding various aspects of the book as given below :

MANUFACTURING SCIENCE (MTU)

BY : Vikas Upadhyay; Vikas Agarwal

Please Fill up the Sheet in Neat and Clean Writing

(i) What topics of your syllabus, important for your examination point of view, are not covered in this book ?

....

....

....

....

(ii) What are the chapters wherein the treatment of the subject matter is not systematic or organised or updated ?

....

....

....

....

(iii) Have you come across misprints/mistakes/factual inaccuracies in the book ? Please specify the chapters and the page numbers.

....

....

....

....

(iv) What other books on the same subject have you found/heard better than the present book ? Please specify in terms of price and quality.

....

....

....

....

(v) Further suggestions and comments for the improvement of this book.

....

....

....

....

(PS : If need be please attach a separate sheet to write your views)

OTHER DETAILS

(i) Who recommended you this book ? Please tick (✓) in the box near the option relevant to you.

☐ Teacher ☐ Friends ☐ Bookseller

....

....

....

....

(ii) Name of the recommending teacher, his designation and address:

....

....

....

....

(iii) Name and address of the bookseller you purchased the book from :

....

....

....

....

(iv) Name and address of your institution (Please mention the University or Board as the case may be)

....

....

....

....

(v) Your name and complete address :

....

....

....

....

(vi) Course you are enrolled in :

....

....

....

....

The best assessment will be awarded each month : The award will be in the form of our publications as decided by the Editorial Board.

Please mail the filled up coupon at your earliest to :

S.K. KATARIA & SONS®

4760-61/23, Ansari Road, Daryaganj, New Delhi-110002

☎ +91-11-23243489, 23269324; Telefax: +91-11-23243489

e-mail: katariabooks@yahoo.com; katariabook@gmail.com

Website : www.skkatariaandsons.com